THIS BOOK BELONGS TO

- -

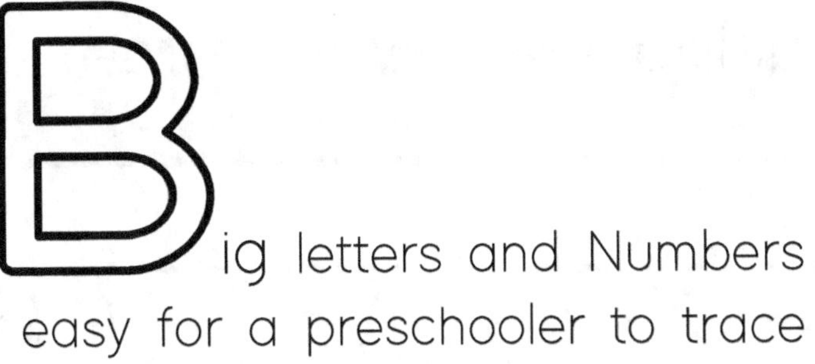

Big letters and Numbers
easy for a preschooler to trace

- Tracing Lines and curves
- Alphabet A-Z in upper and lowercase
- Numbers 1-10

Start at the circle and trace the dotted line.
End at the arrow.

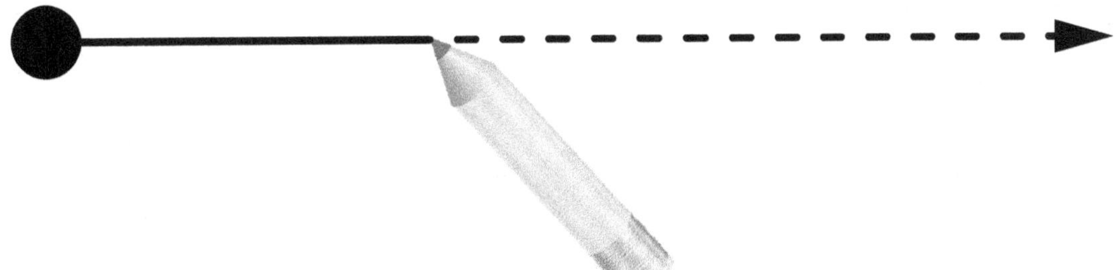

Trace the dotted line step by step. Finally let your kids write the letter by themselves without dotted guide in the Excellent box.

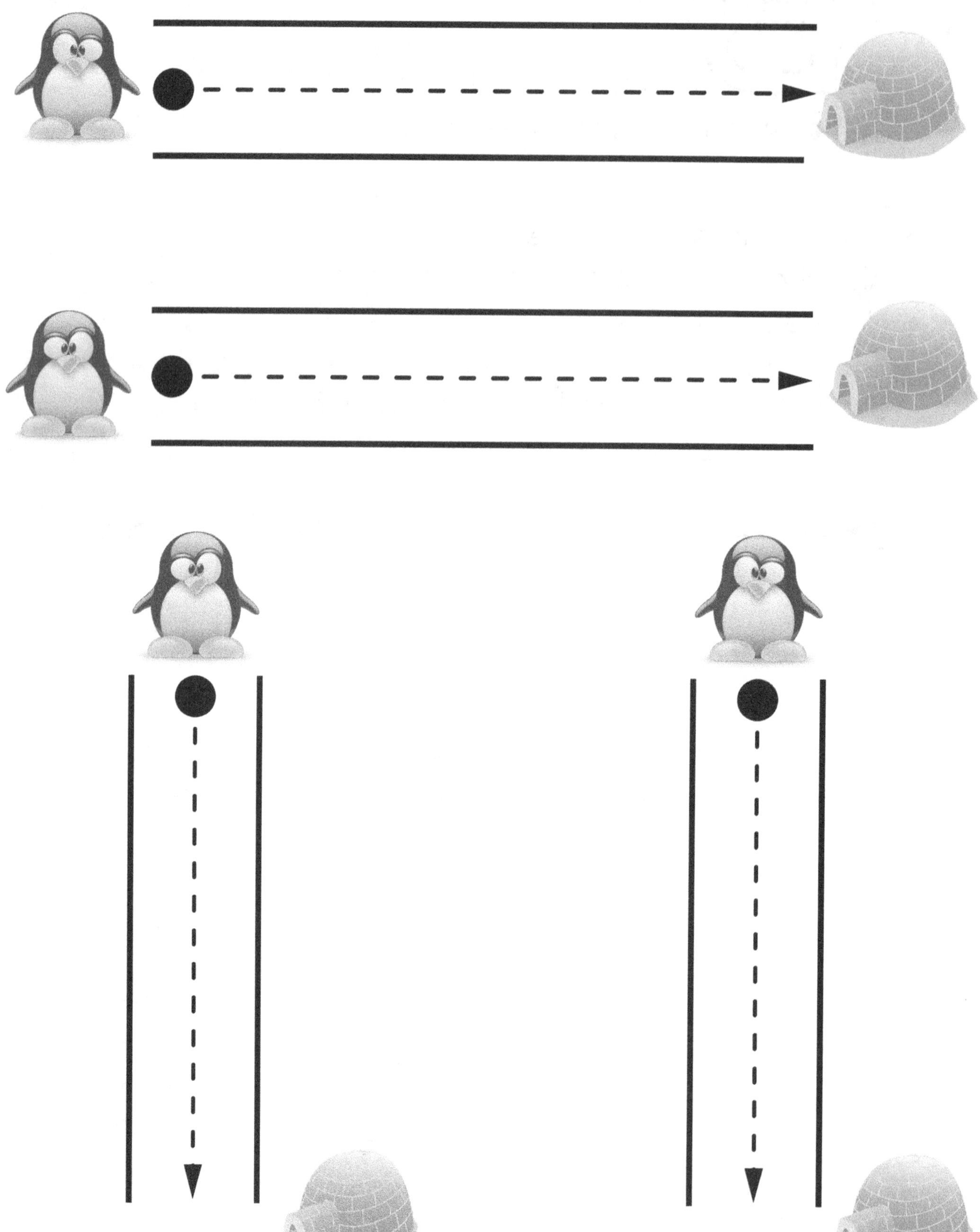

HELP Penguins get home
by tracing their way home.

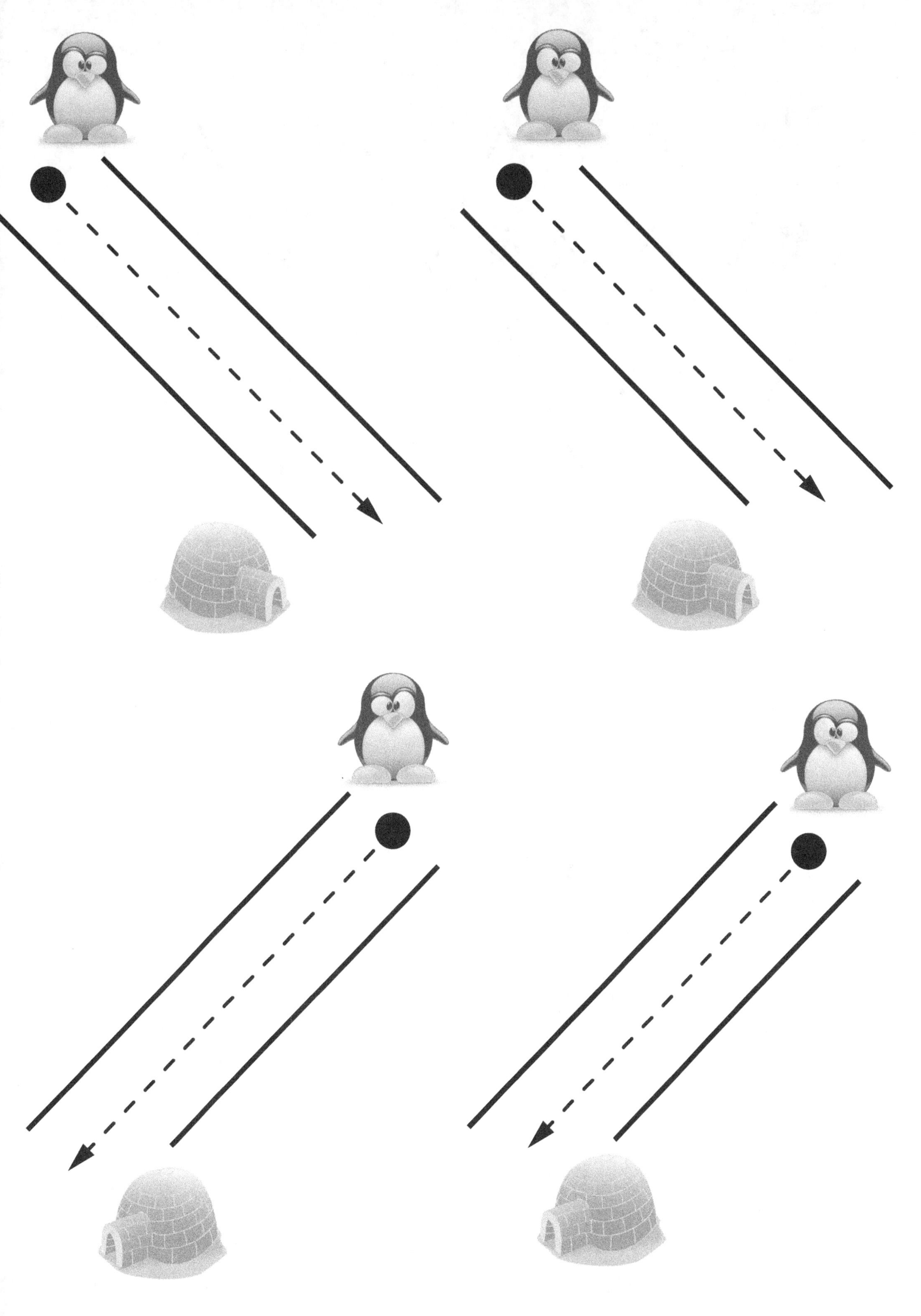

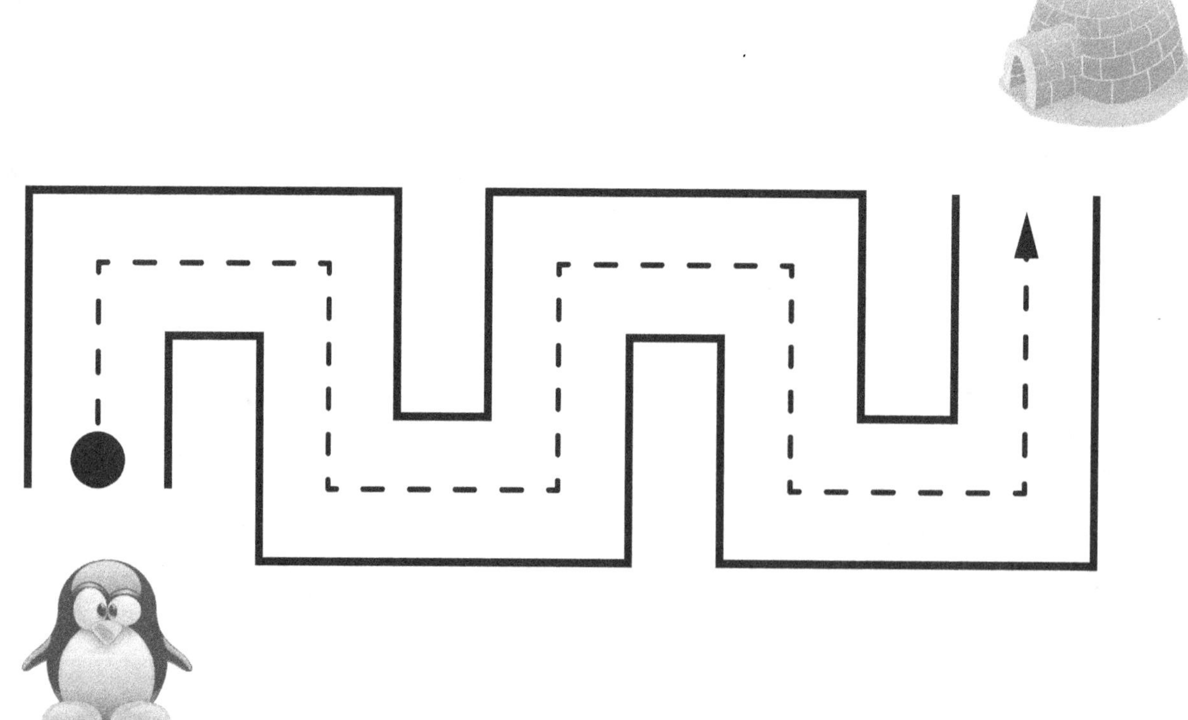

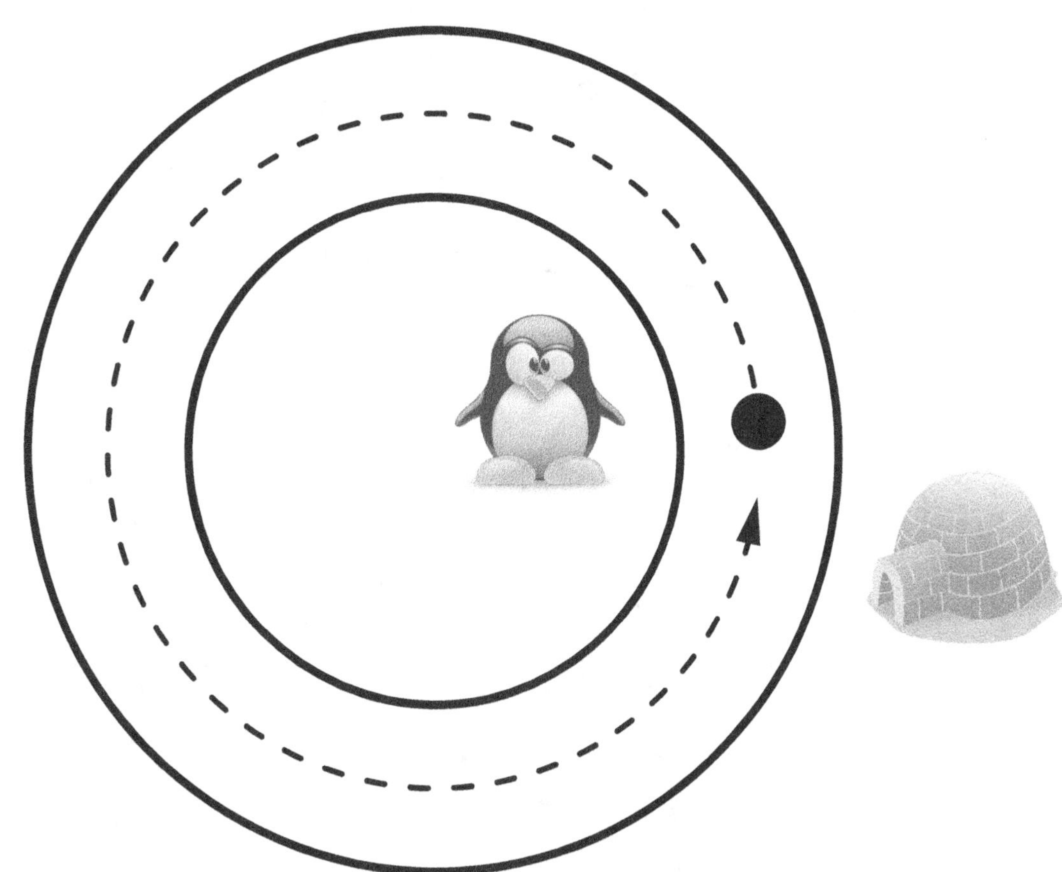

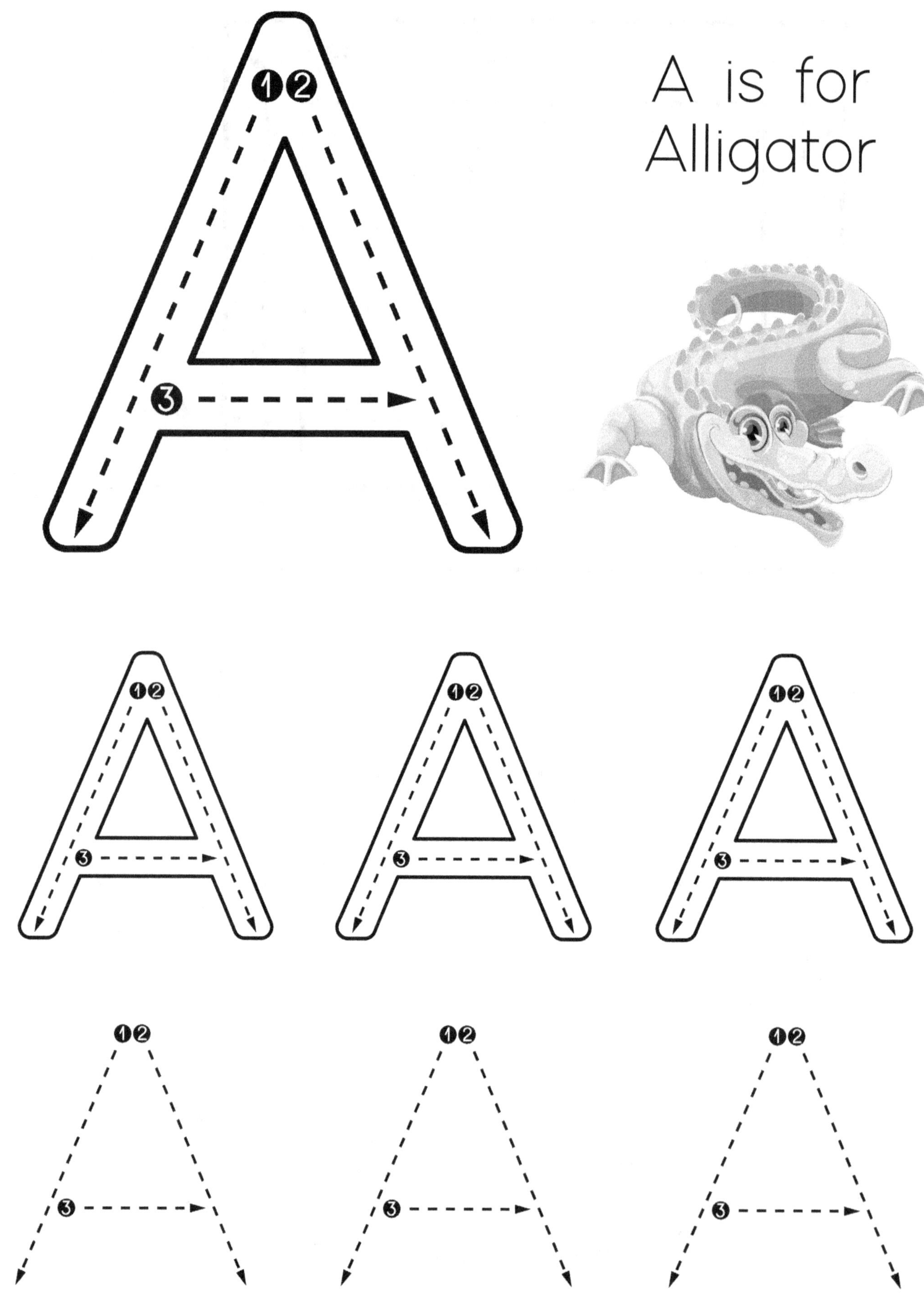

A is for
Alligator

A A A

Excellent!
Now
you can write

A

B is for
Bananas

Excellent!
Now
you can write

B

C is for
Cat

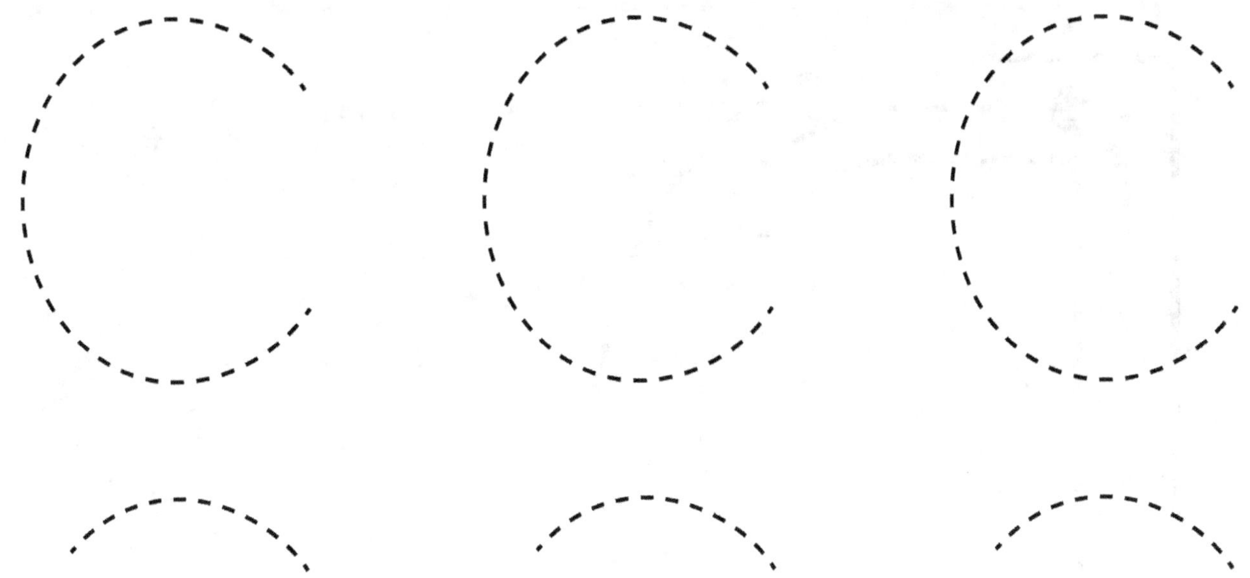

Excellent!
Now
you can write

C

D is for
Dog

Excellent!
Now
you can write

D

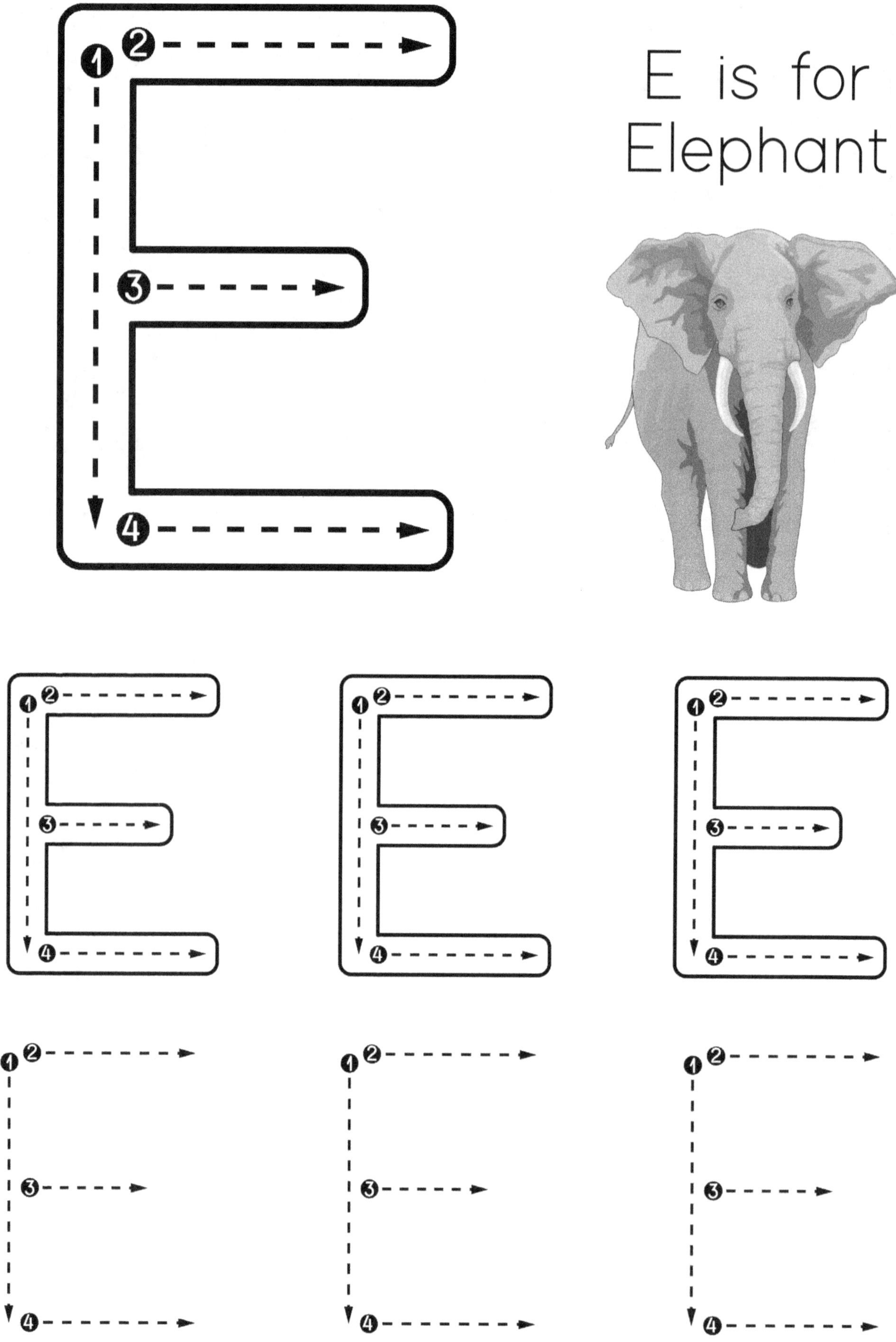

E is for
Elephant

Excellent!
Now
you can write

E

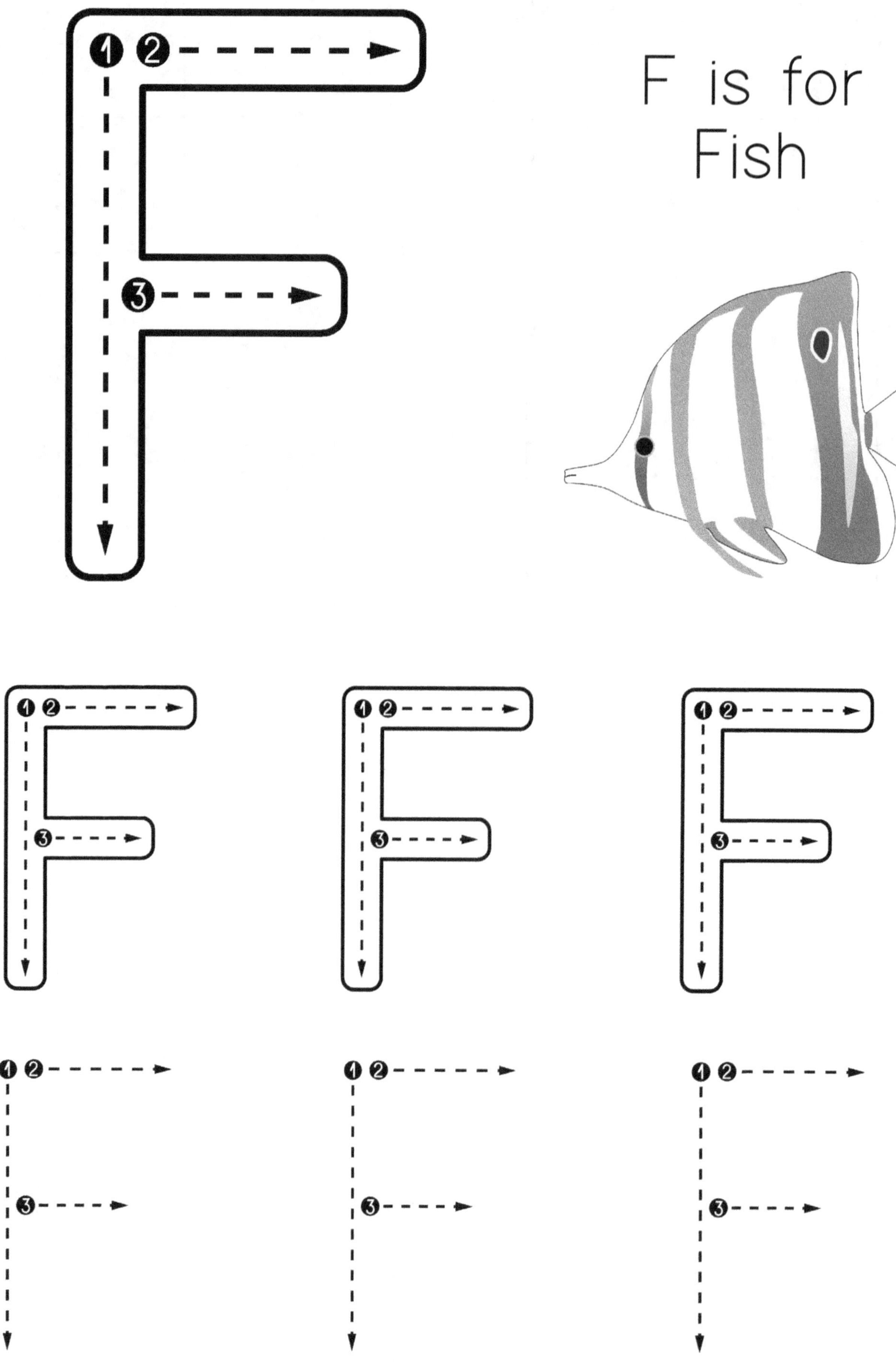

F is for
Fish

Excellent!
Now
you can write

F

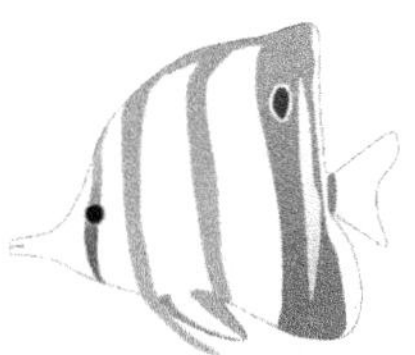

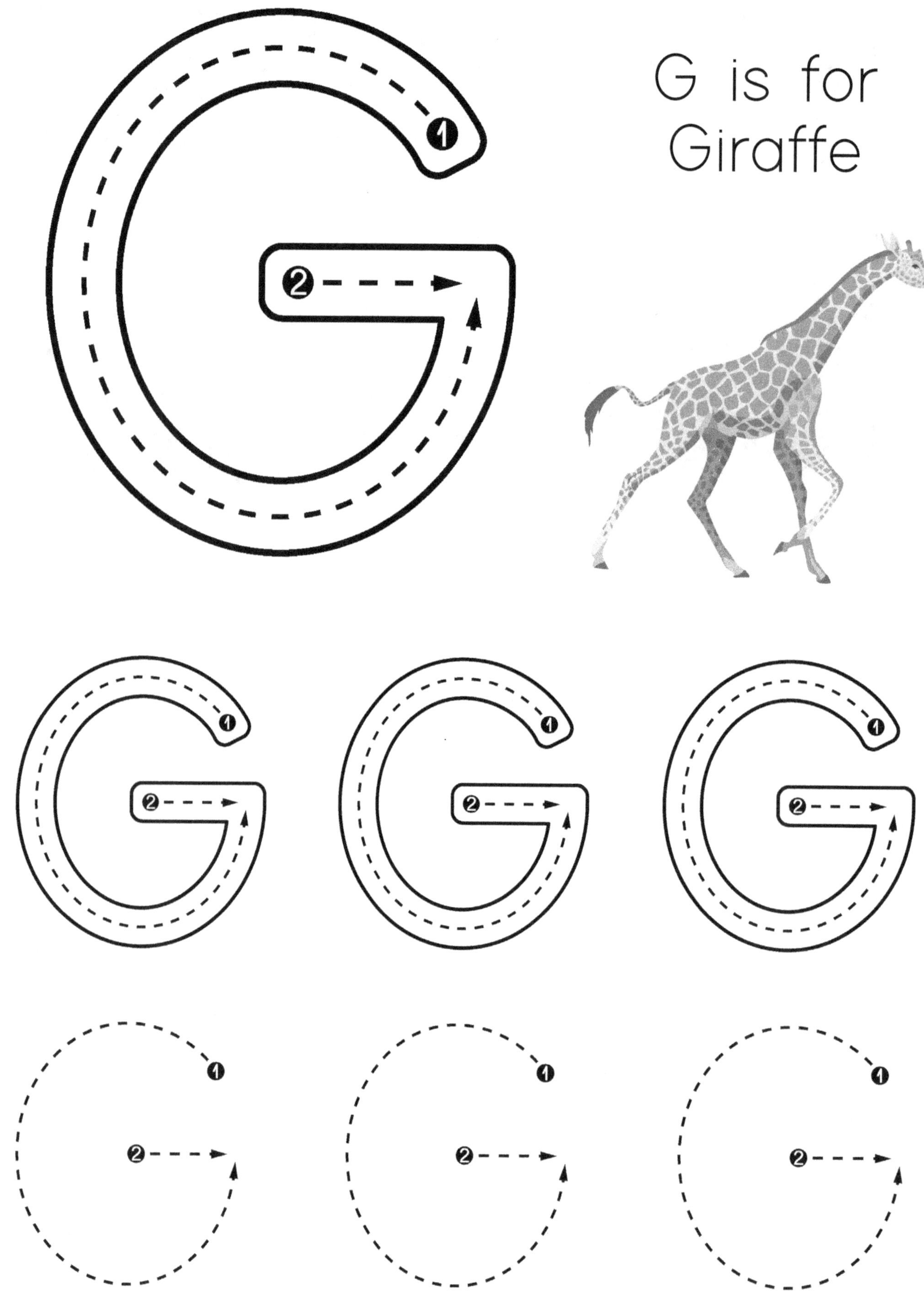

G is for
Giraffe

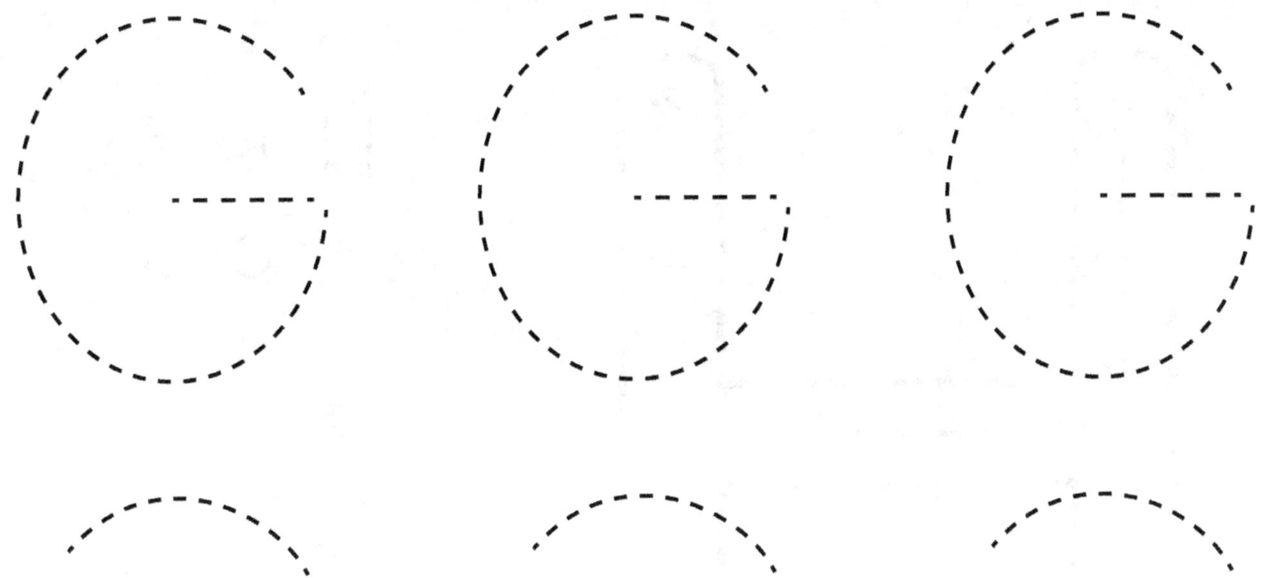

Excellent!
Now
you can write
G

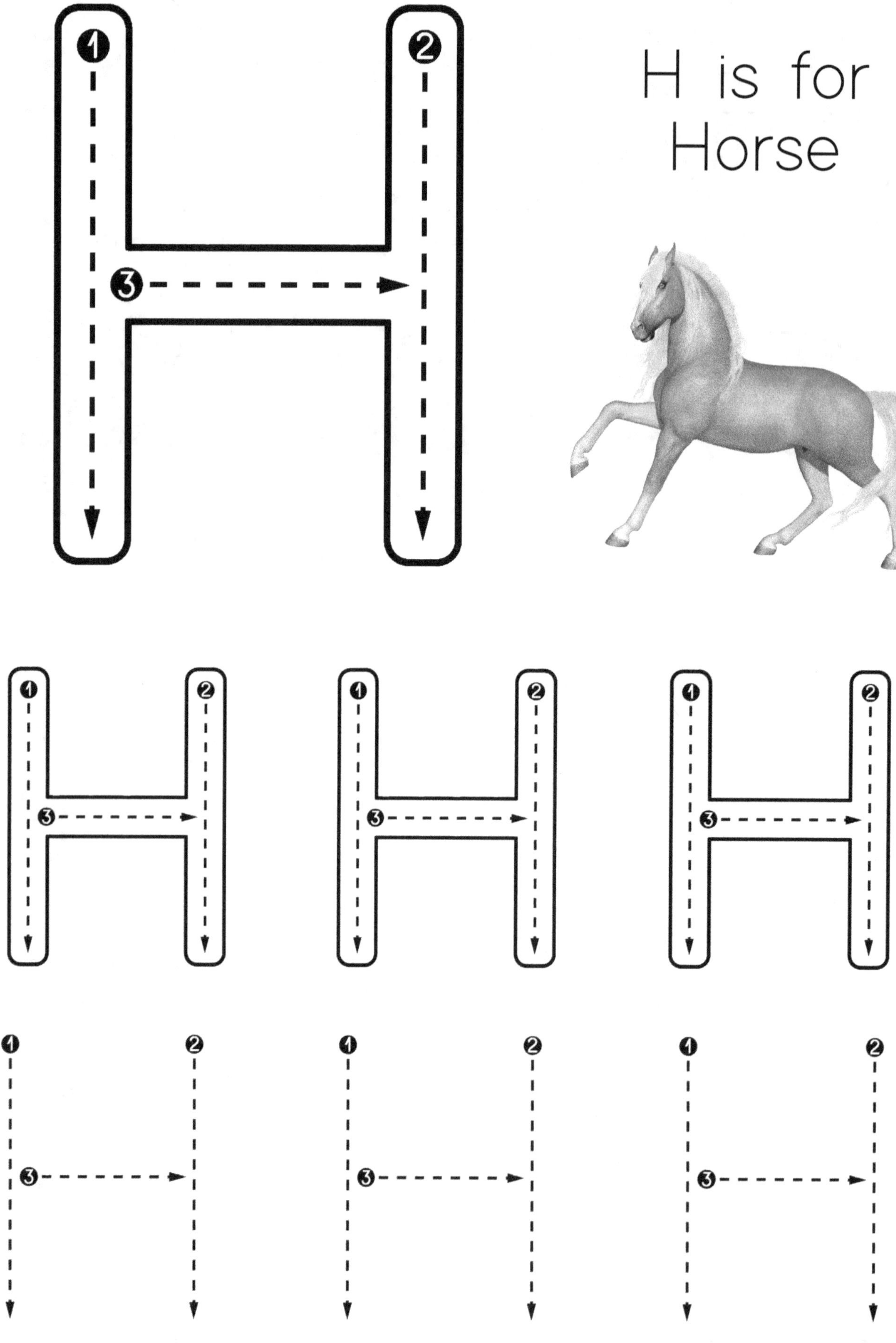

H is for
Horse

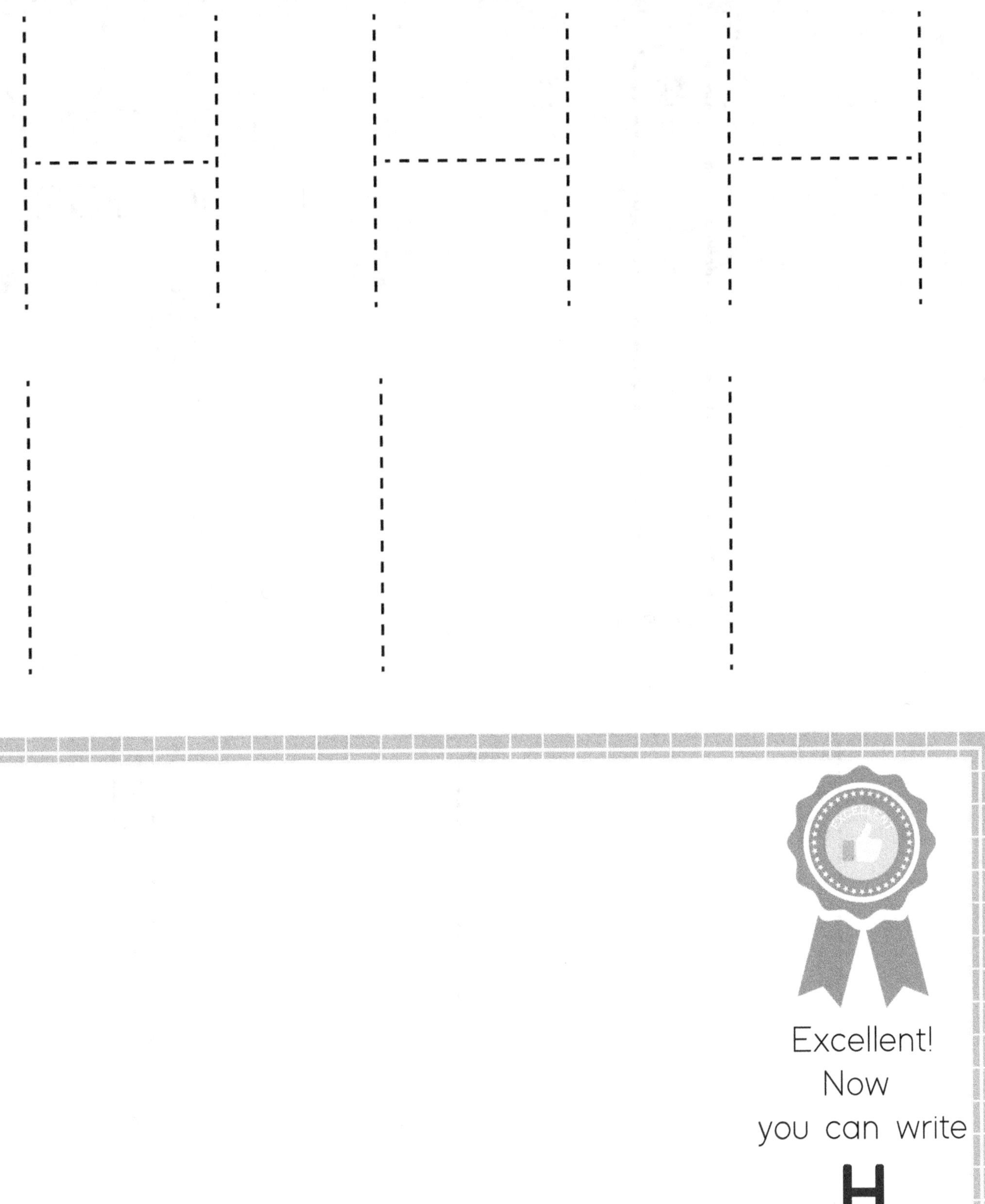
Excellent!
Now
you can write

H

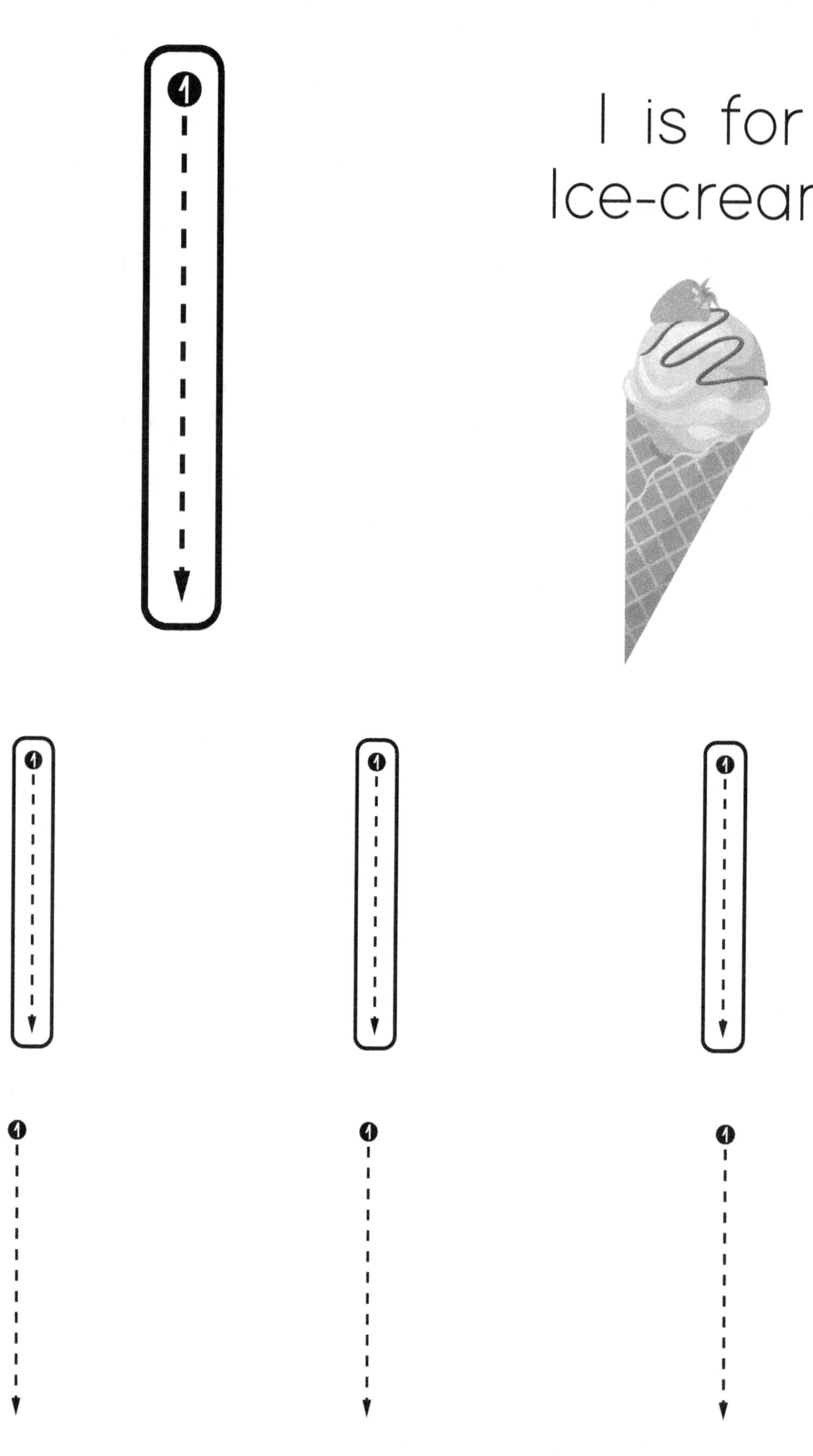
I is for
Ice-cream

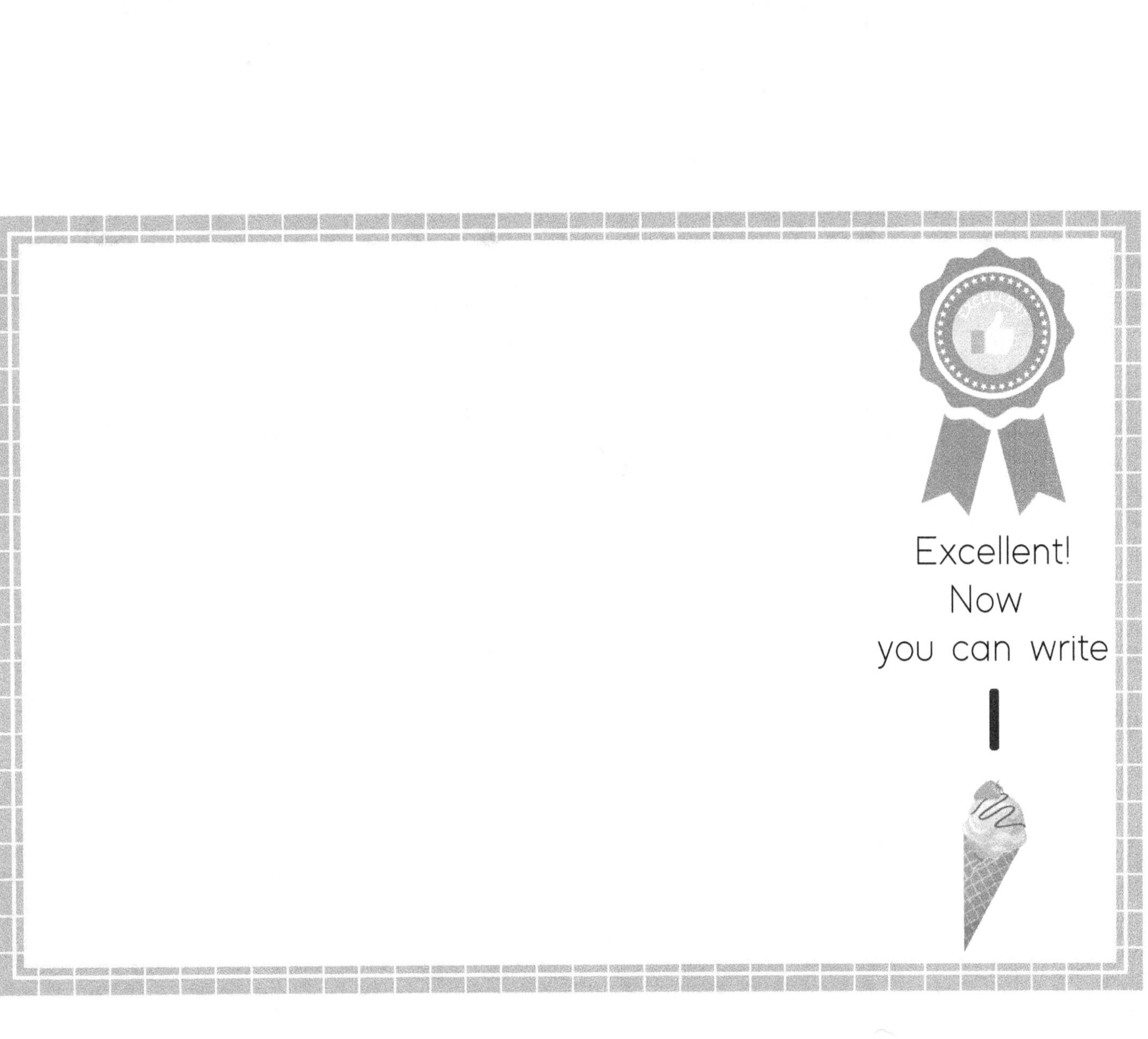
Excellent!
Now
you can write
I

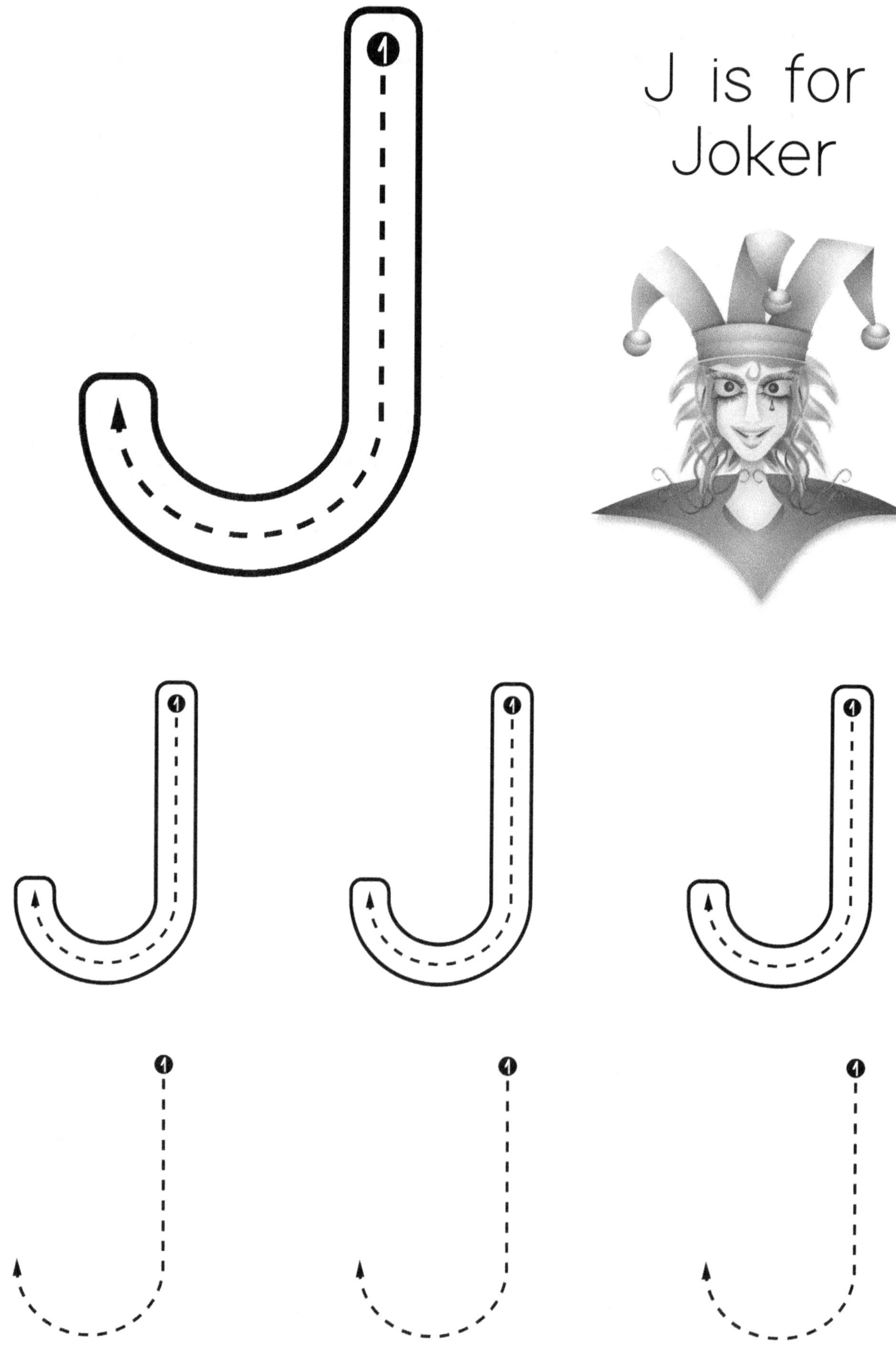

J is for
Joker

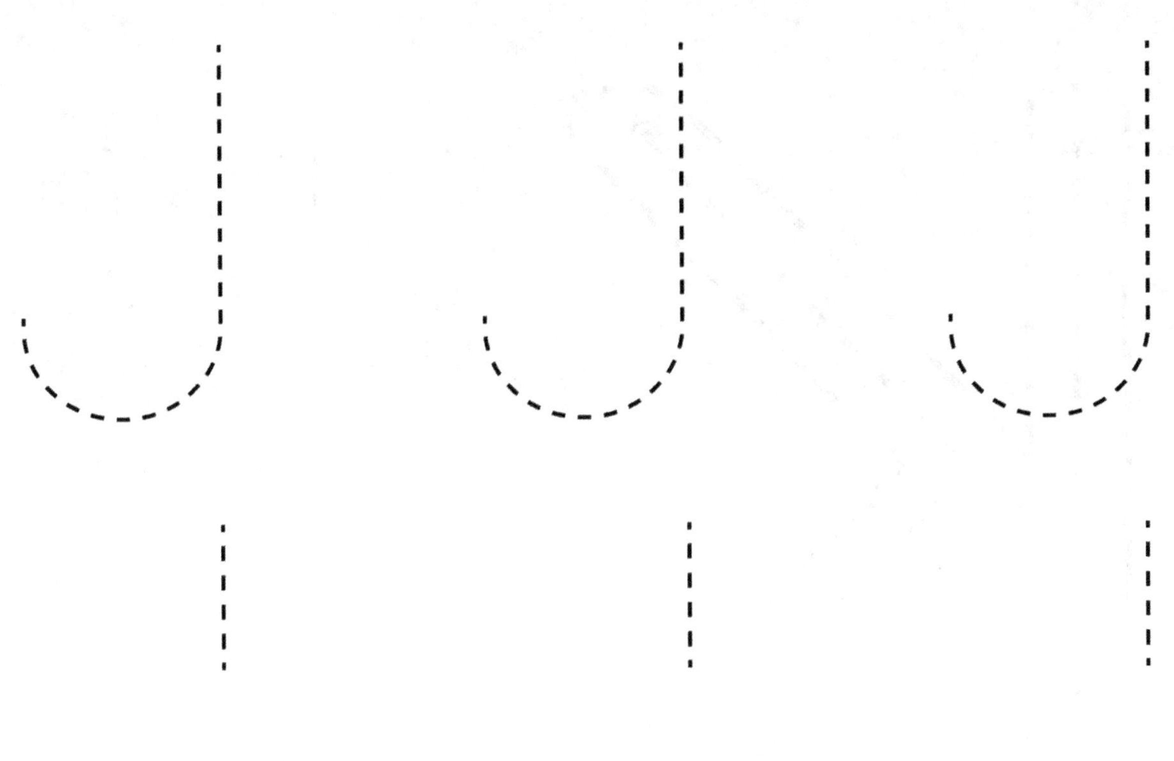

Excellent!
Now
you can write

J

K is for
Kite

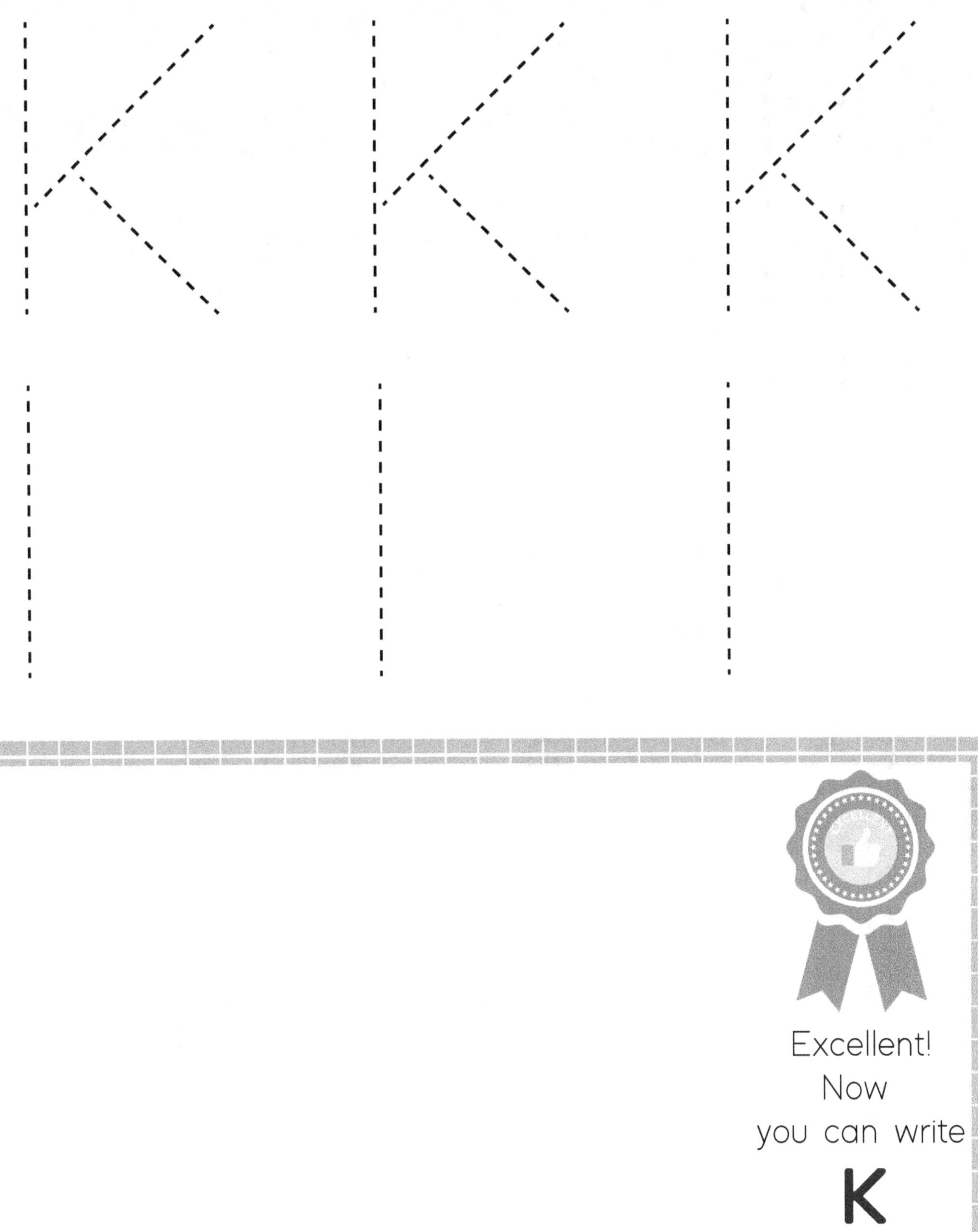

Excellent!
Now
you can write
K

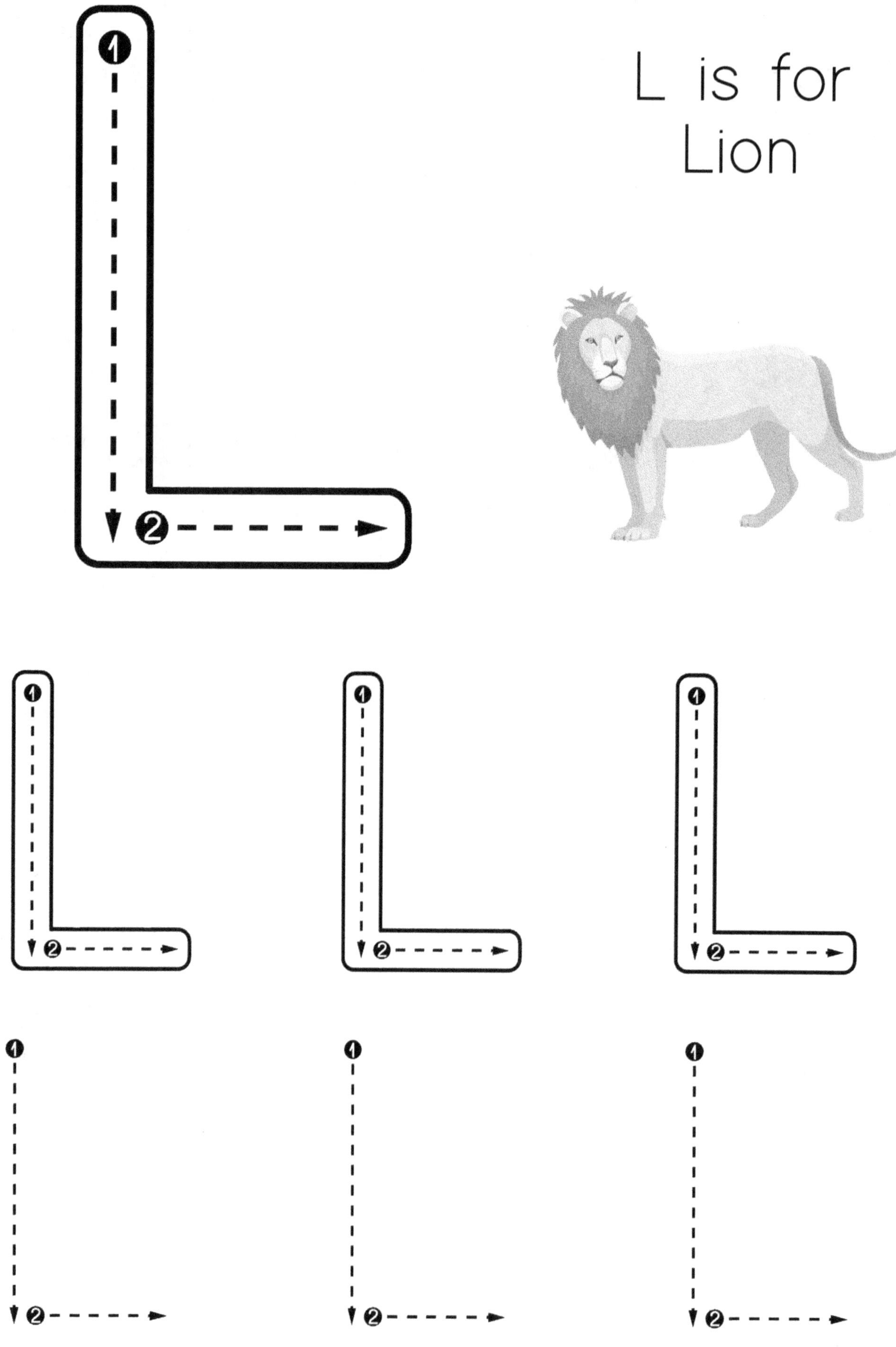

L is for
Lion

Excellent!
Now
you can write

L

M is for
Mango

Excellent!
Now
you can write

M

N is for
Night

Excellent!
Now
you can write

N

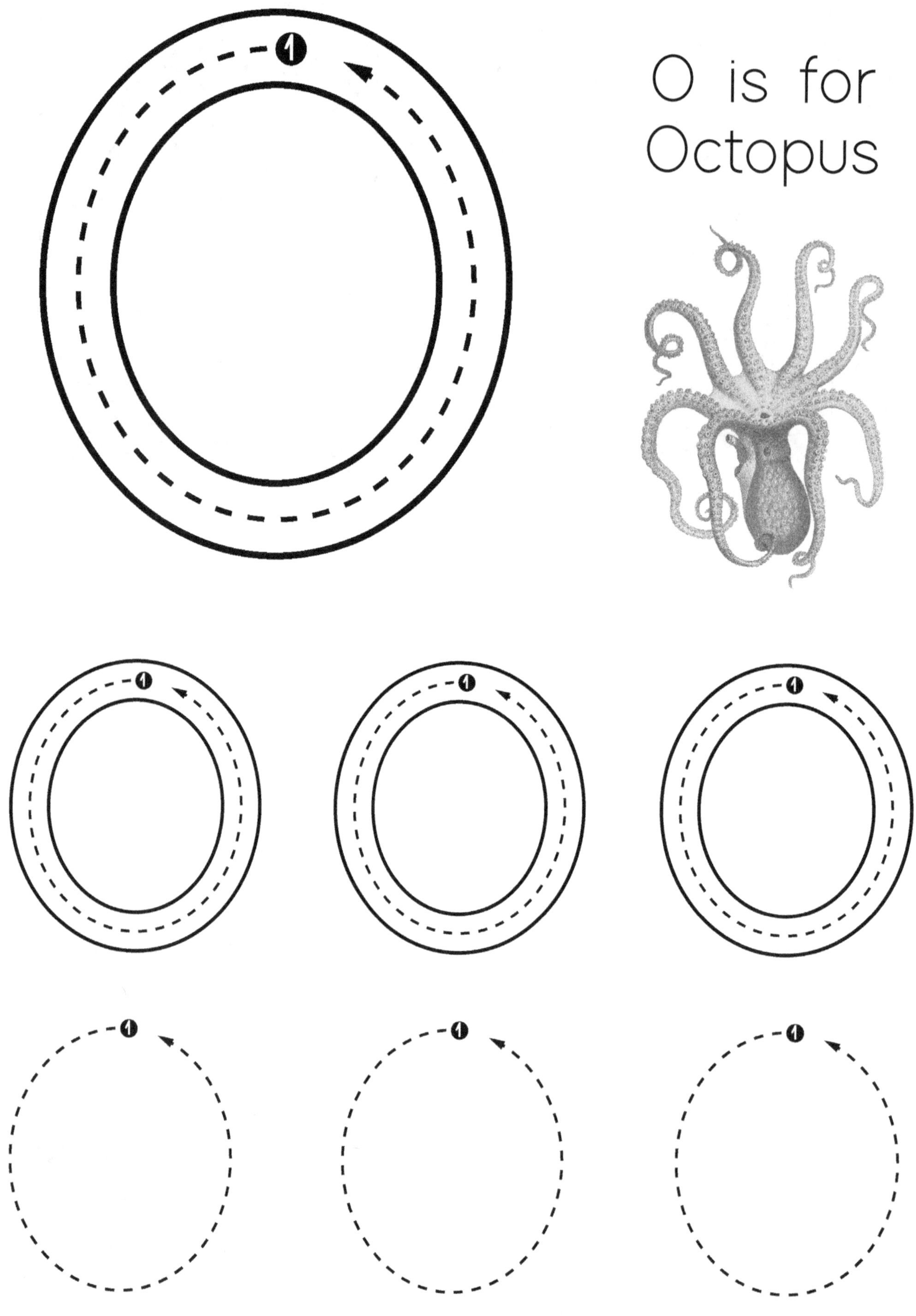

O is for
Octopus

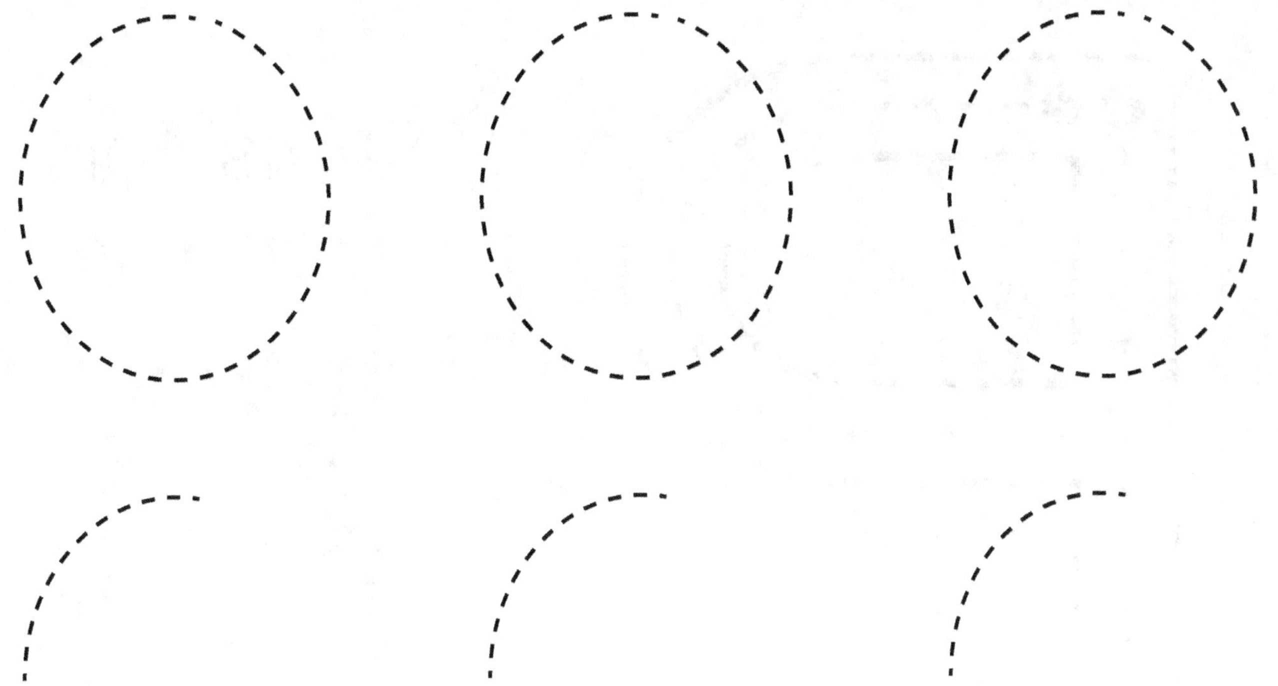

Excellent!
Now
you can write

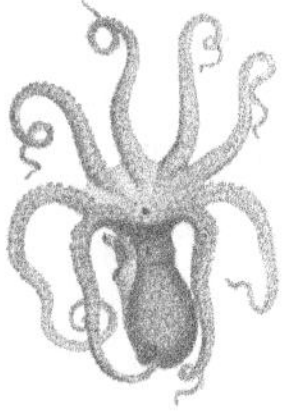

❶❷
P is for
Penguin

P P P

Excellent!
Now
you can write

P

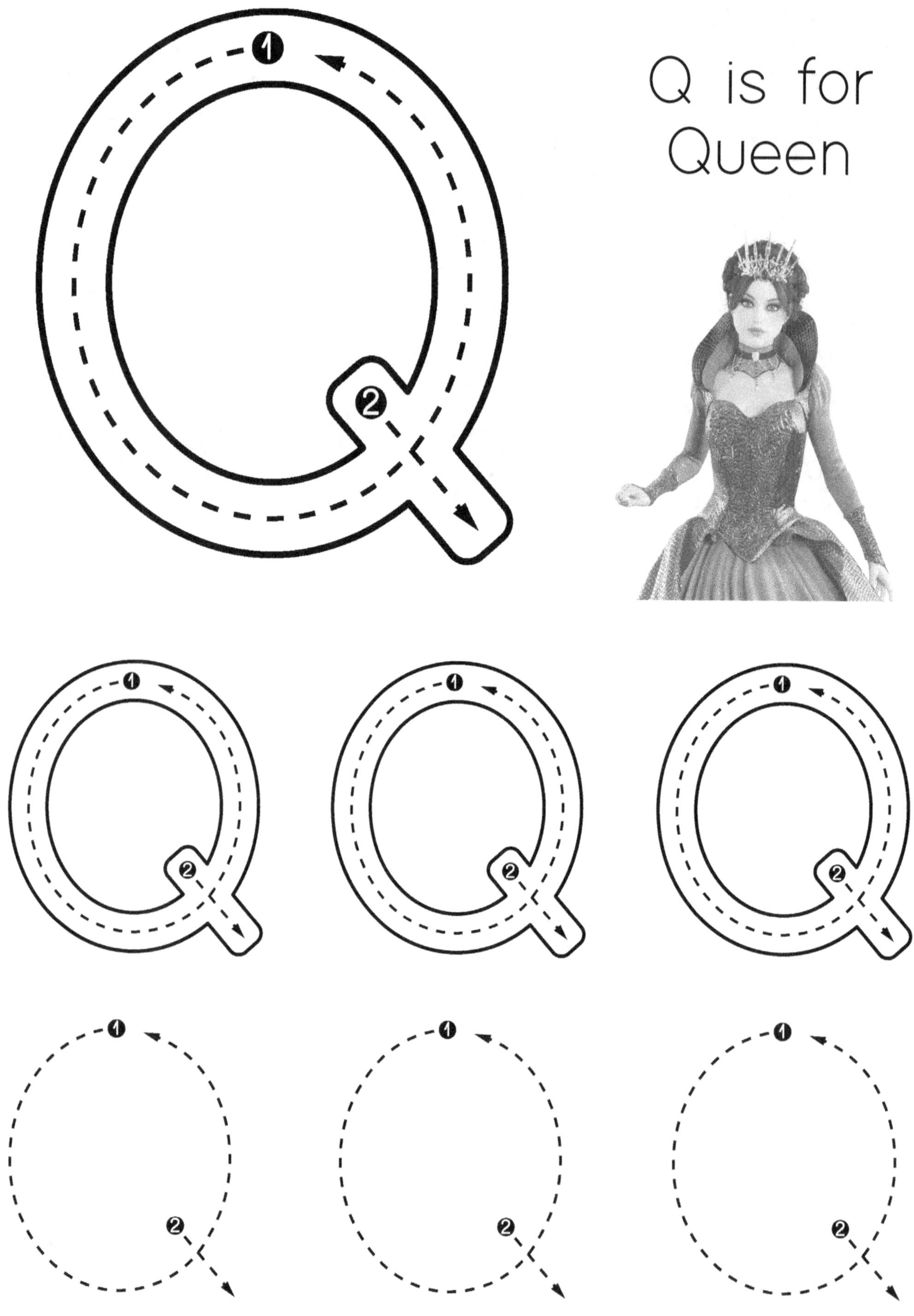

1
2
Q is for
Queen

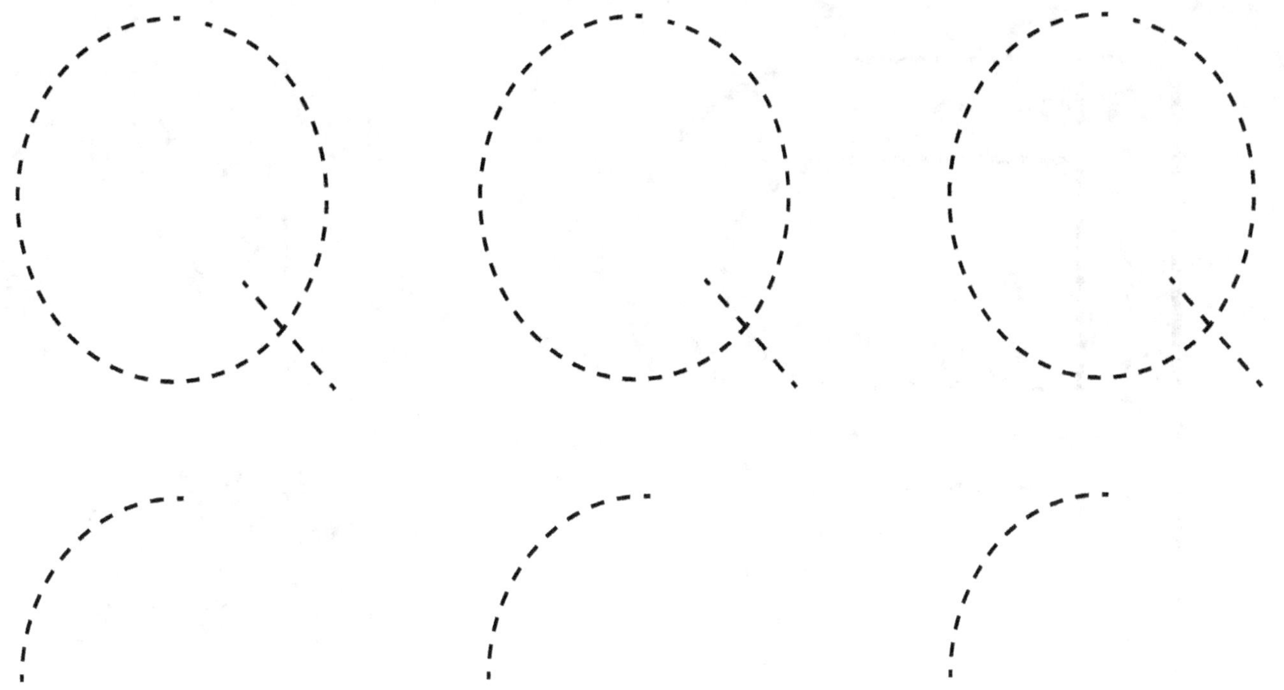

Excellent!
Now
you can write

Q

R is for
Rat

Excellent!
Now
you can write

R

S is for
Sheep

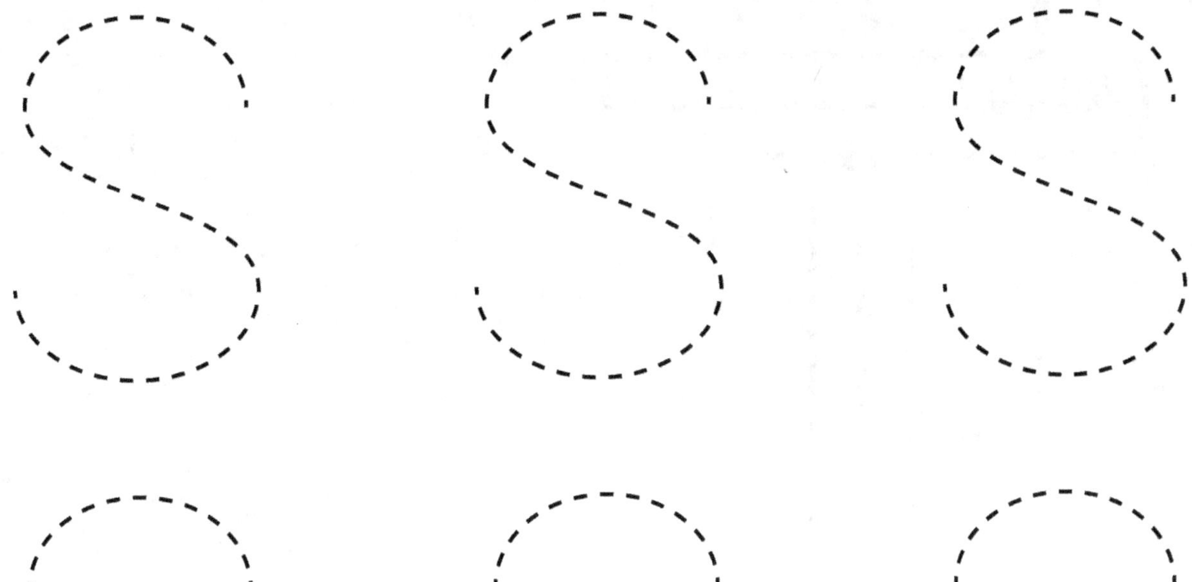

Excellent!
Now
you can write

S

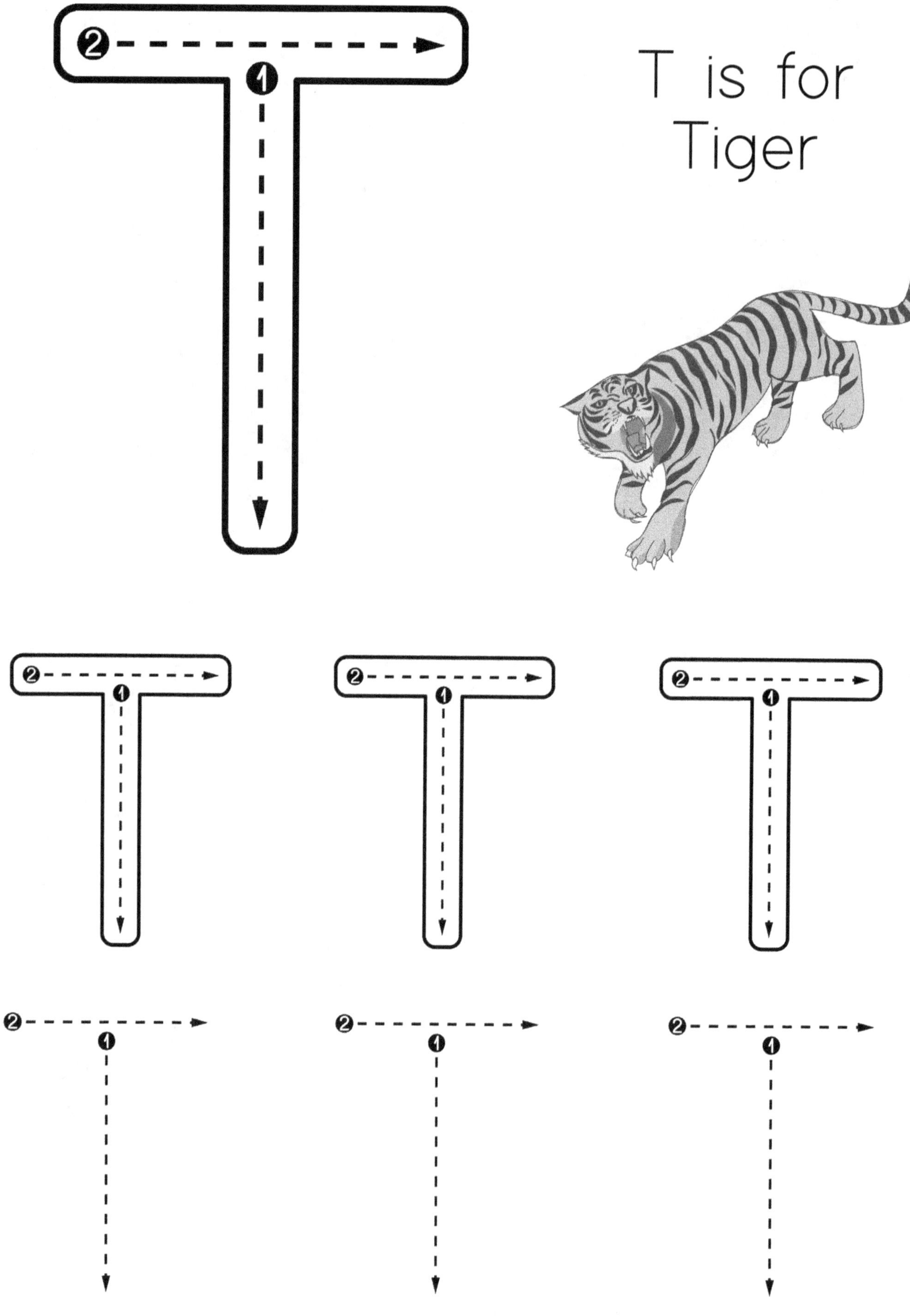

T is for
Tiger

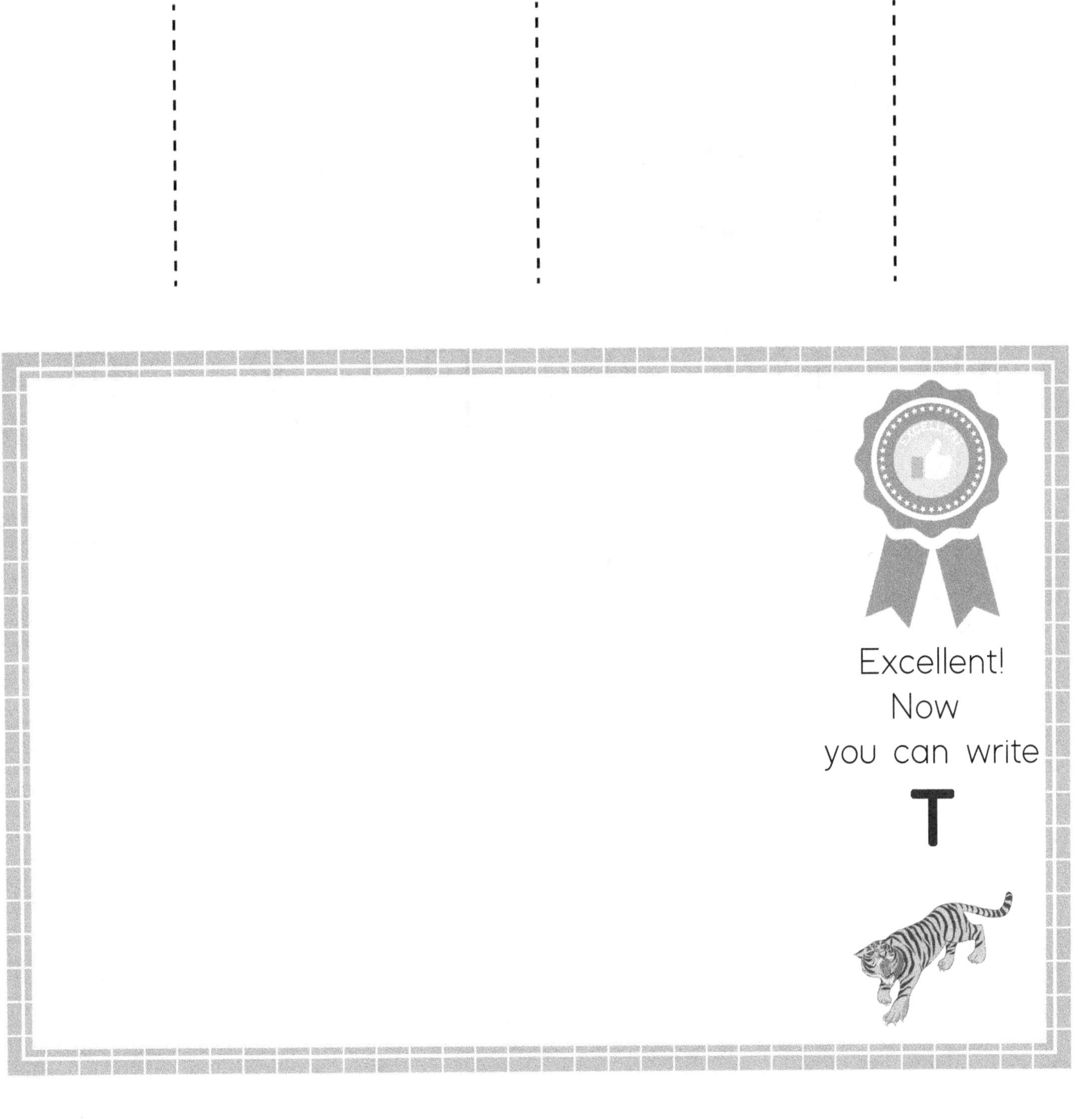

Excellent!
Now
you can write
T

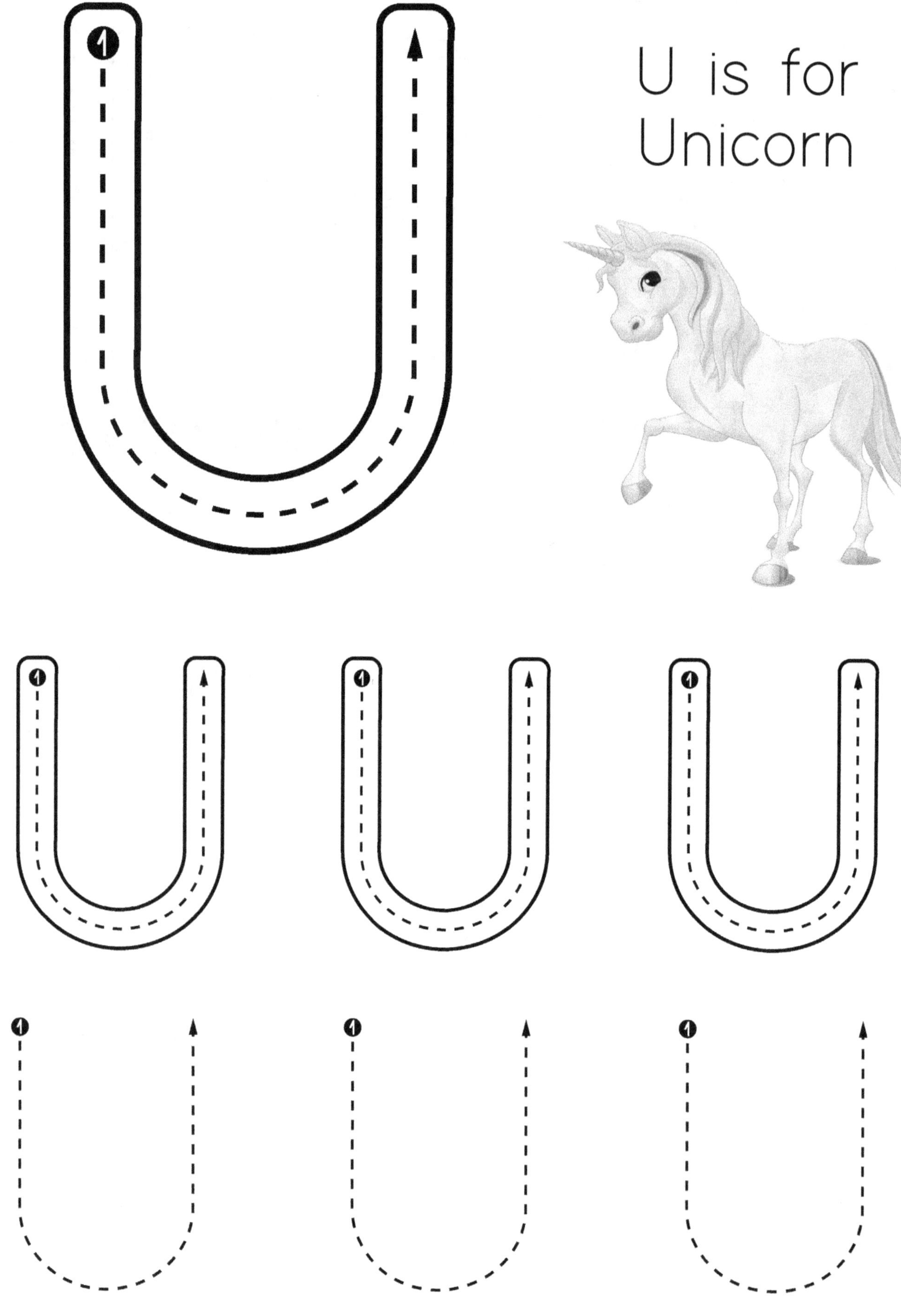

U is for
Unicorn

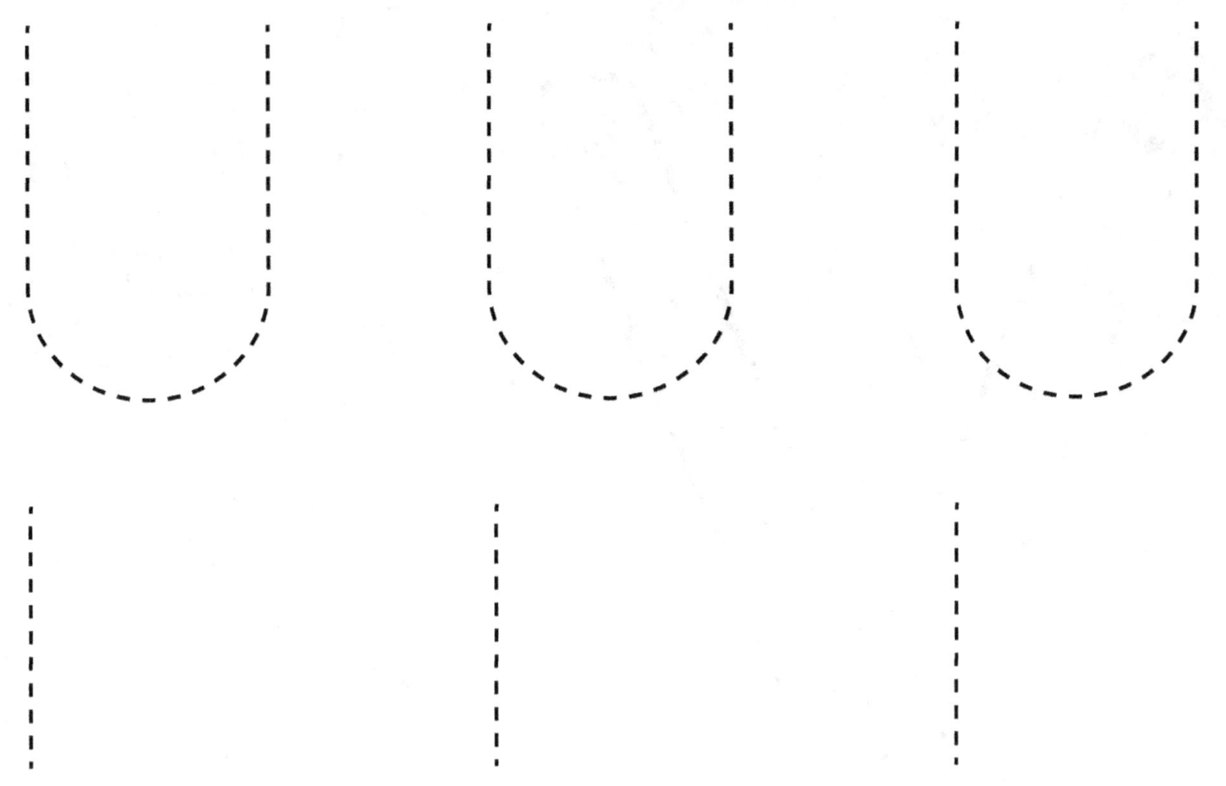

Excellent!
Now
you can write
U

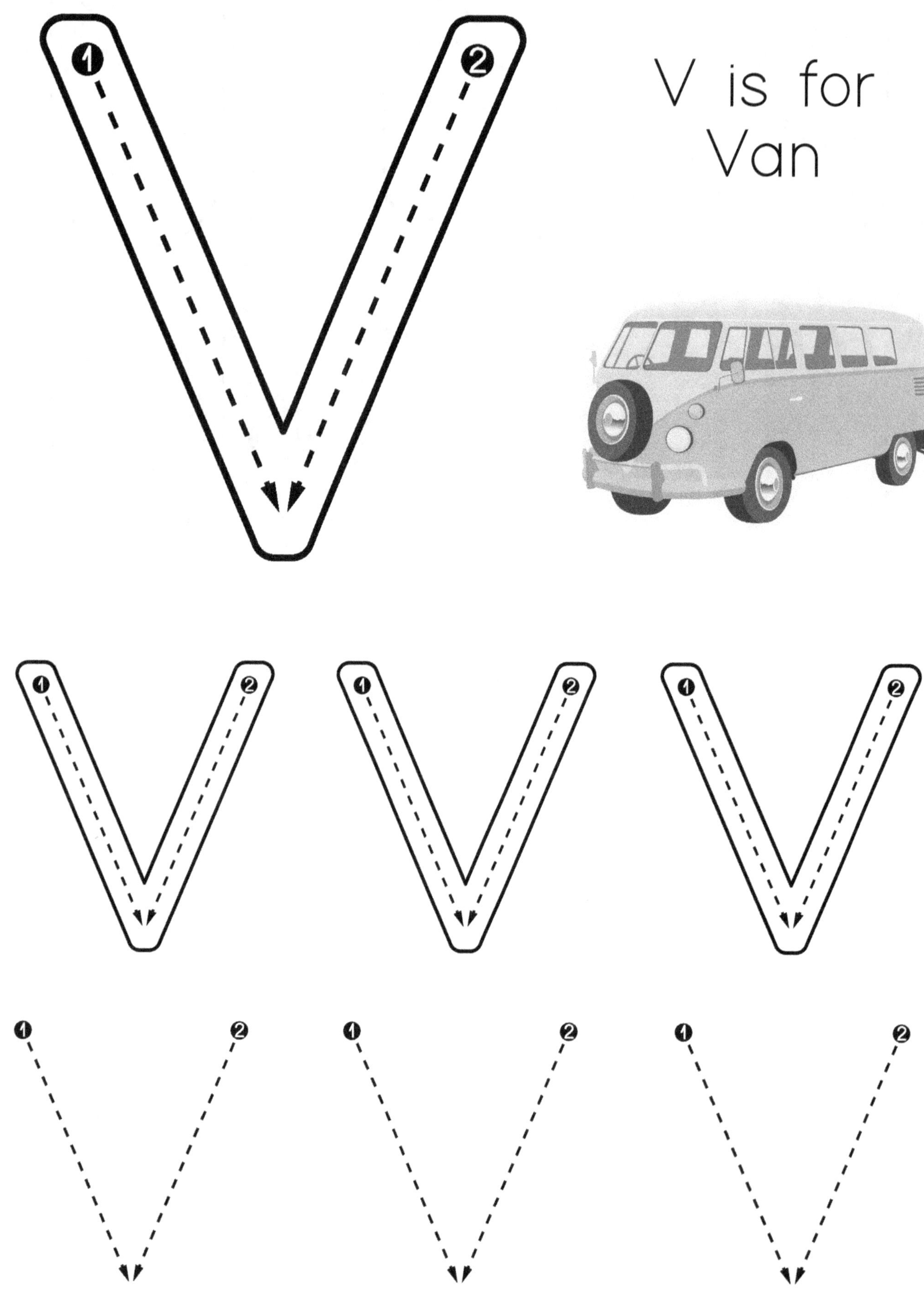
V is for
Van

Excellent!
Now
you can write

V

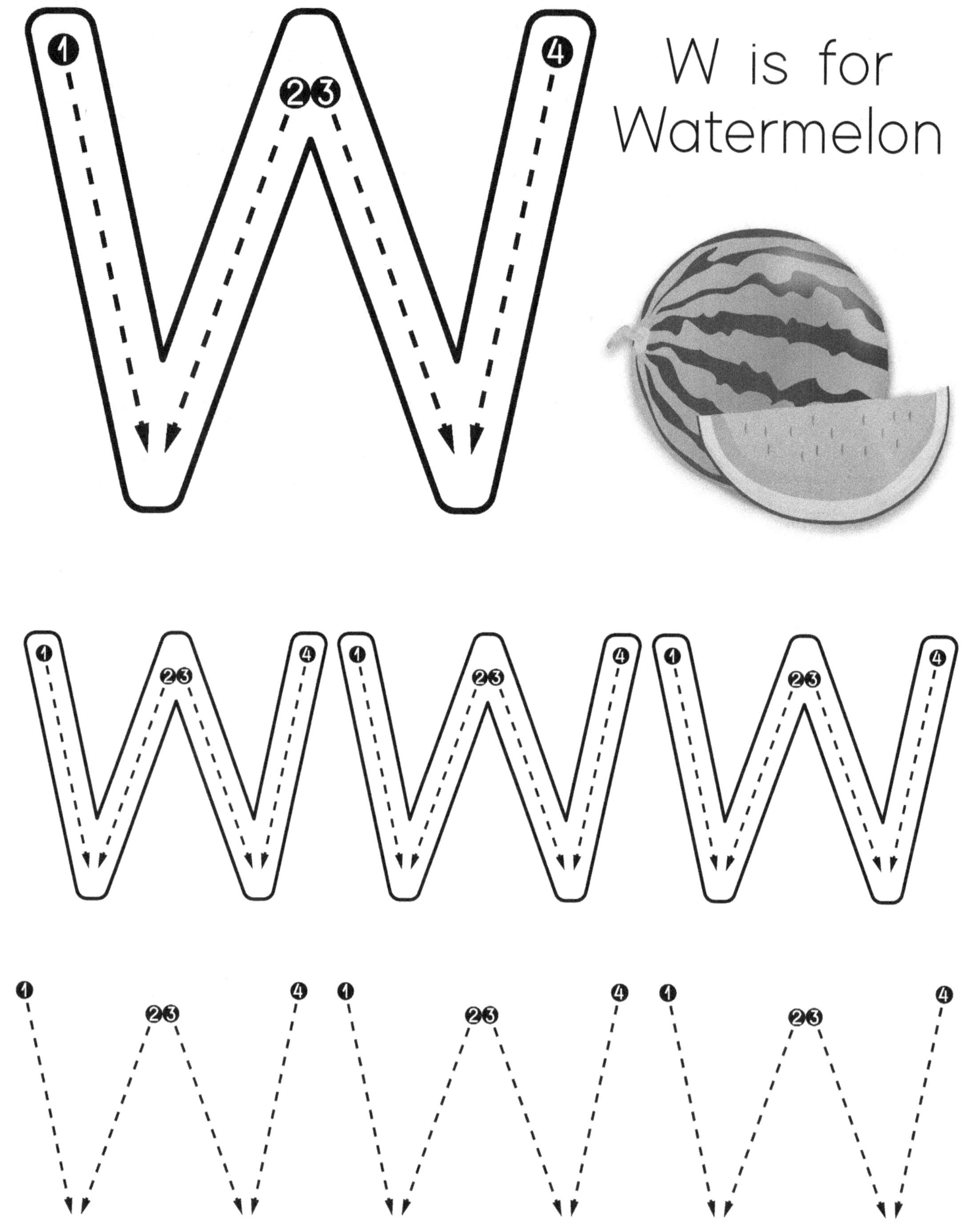

W is for
Watermelon

Excellent!
Now
you can write

W

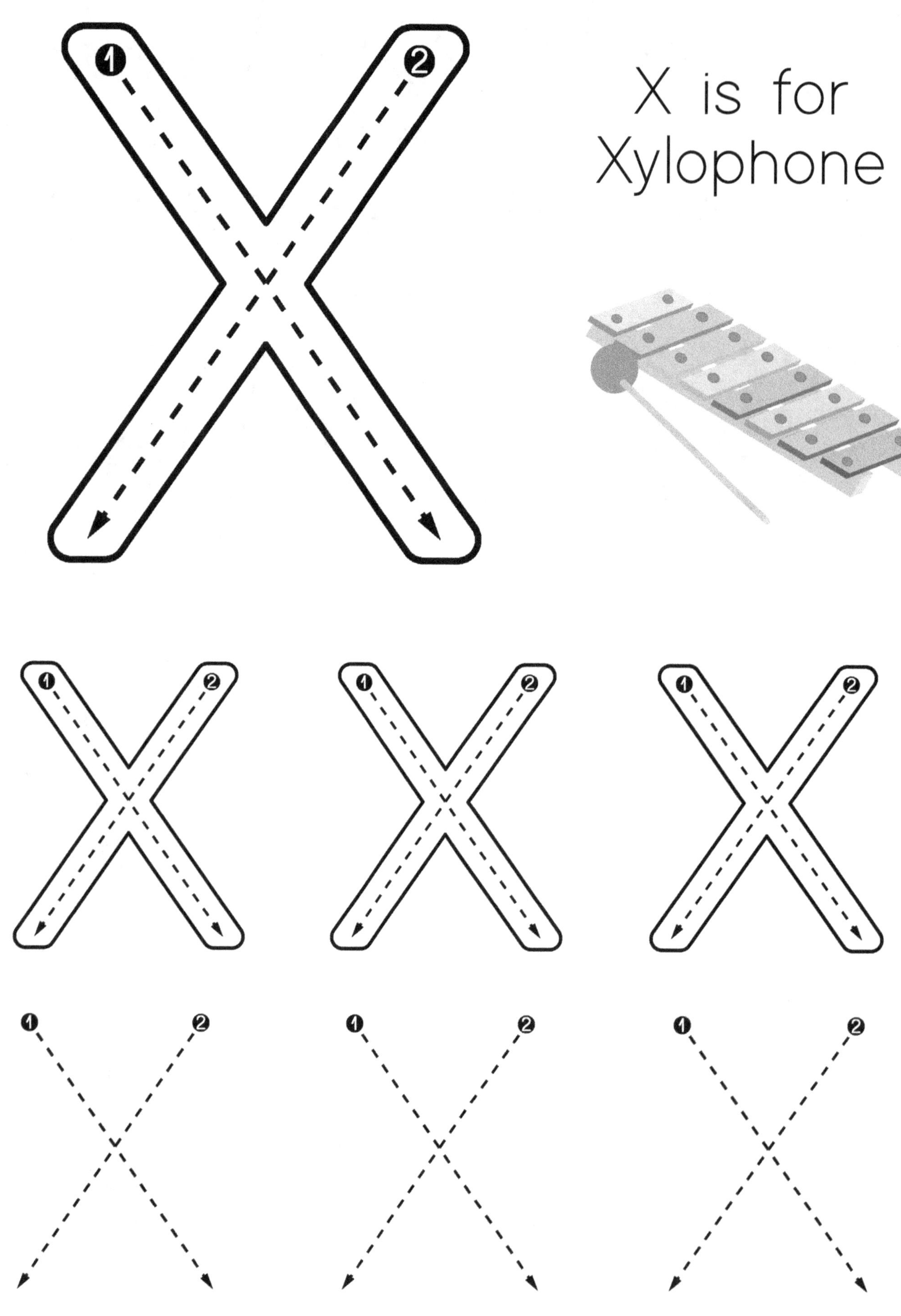
X is for
Xylophone

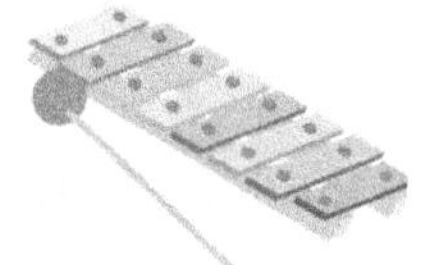

Excellent!
Now
you can write

X

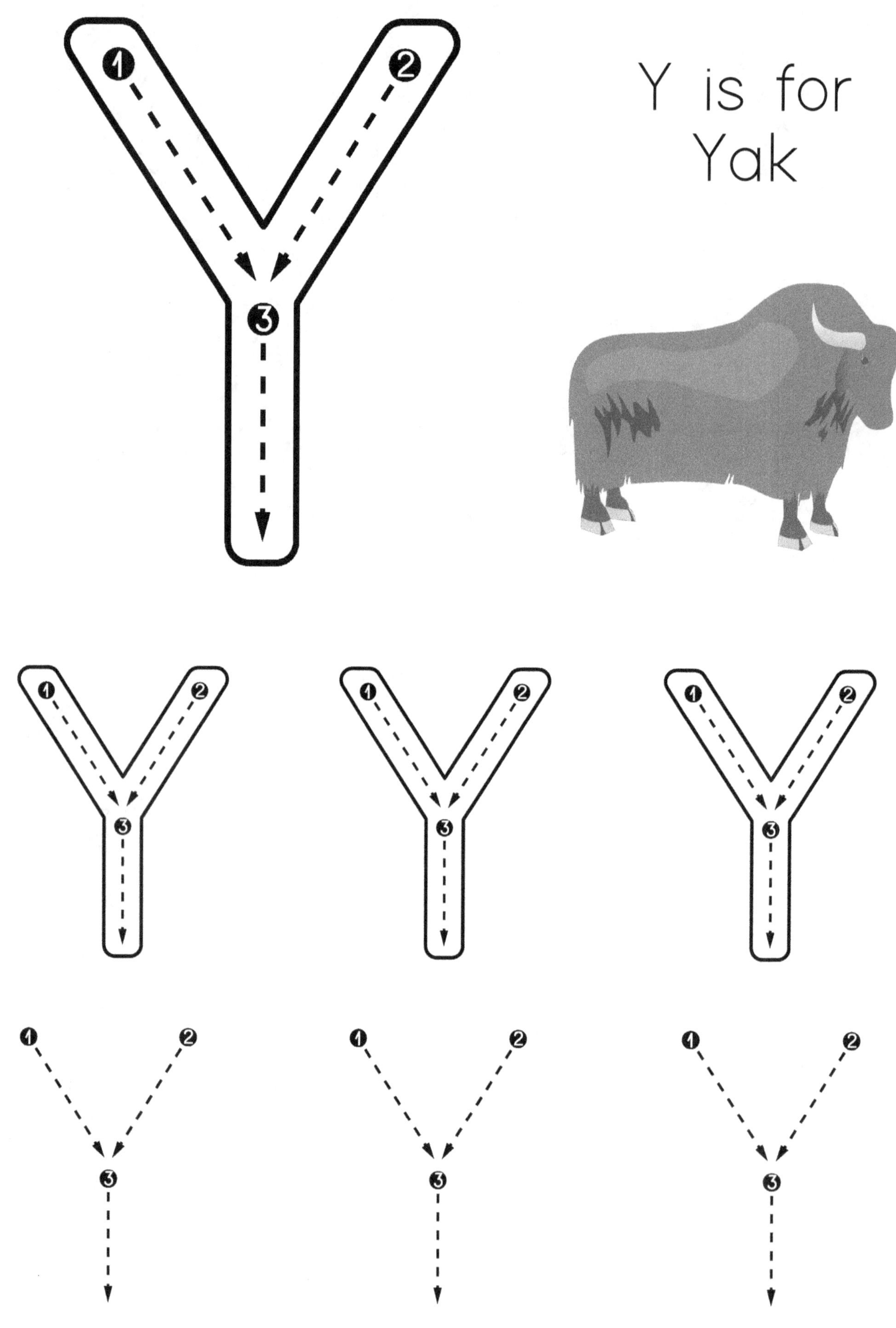

Y is for
Yak

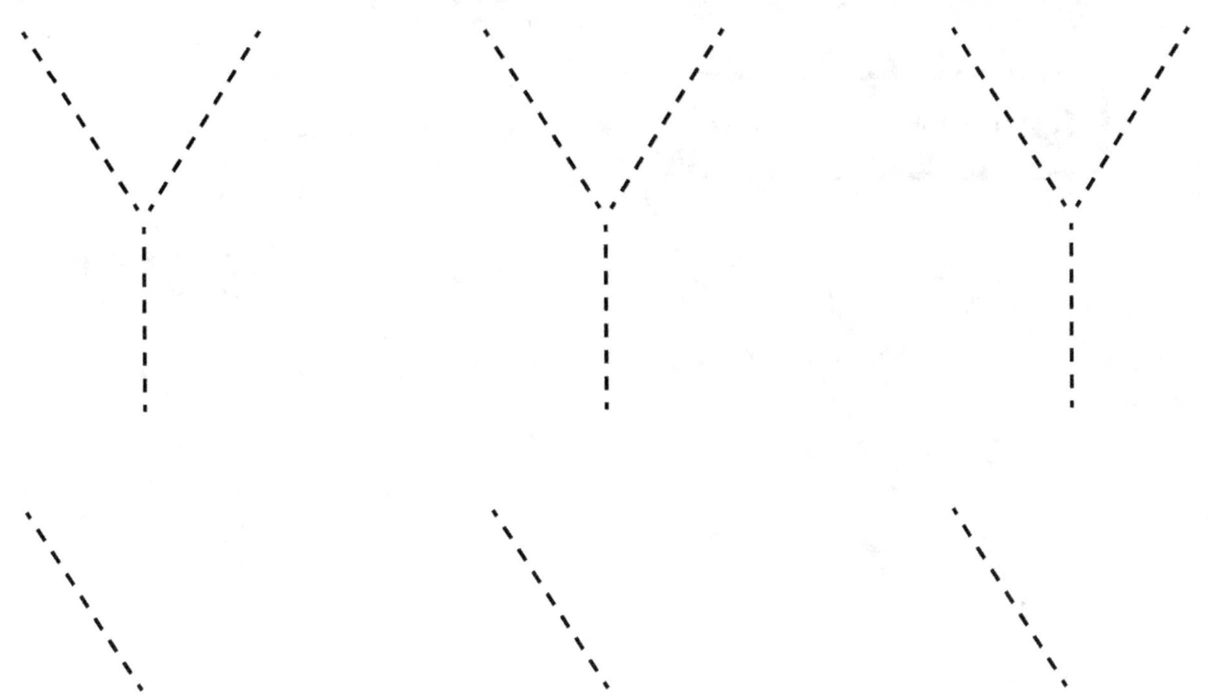

Excellent!
Now
you can write

Y

Z is for
Zebra

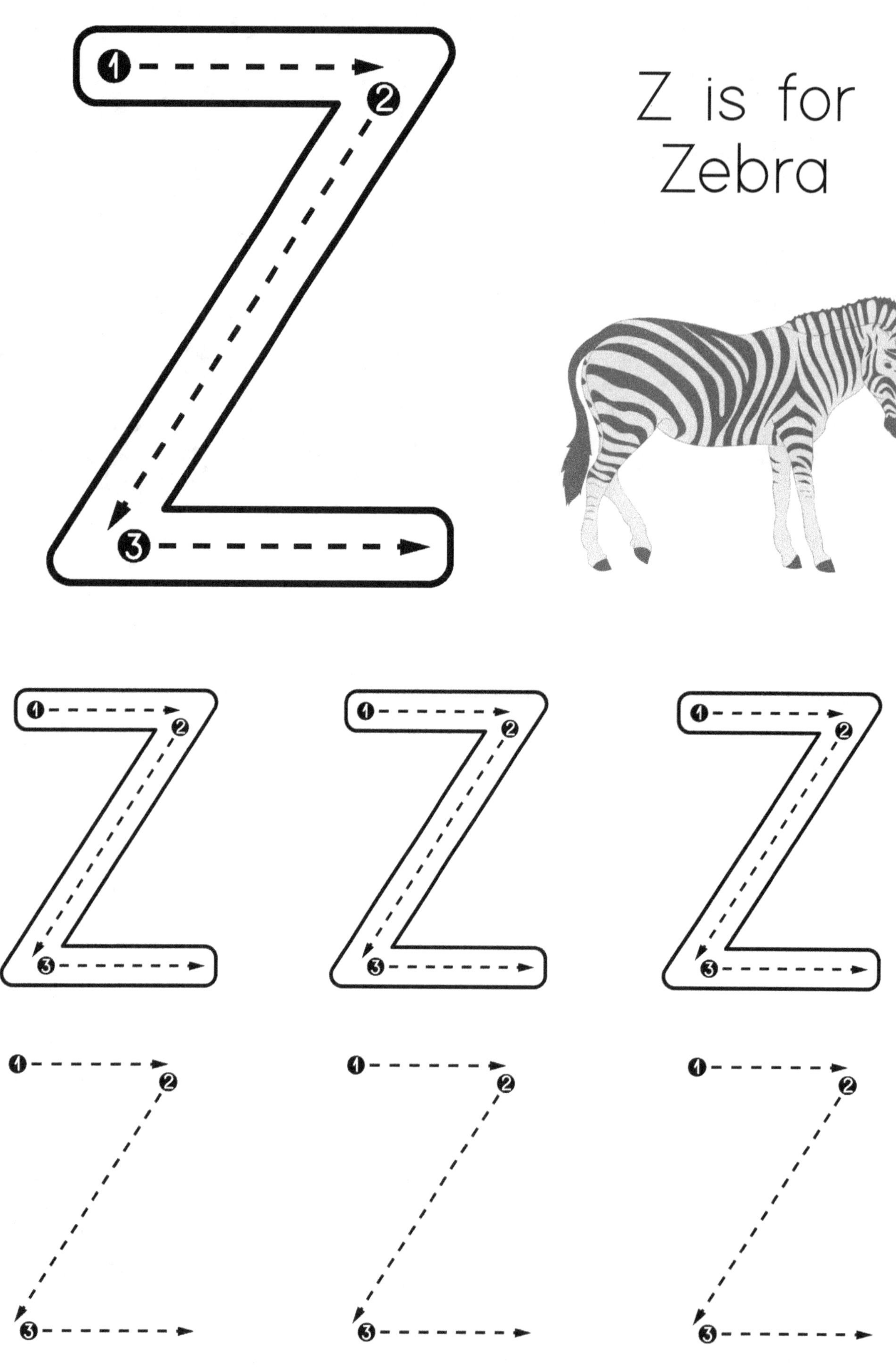

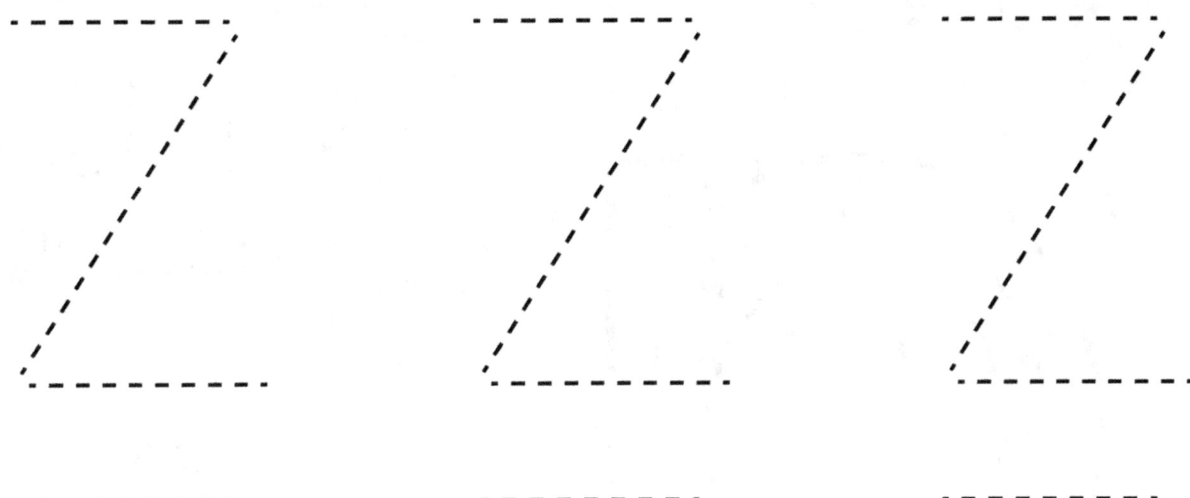

Excellent!
Now
you can write

Z

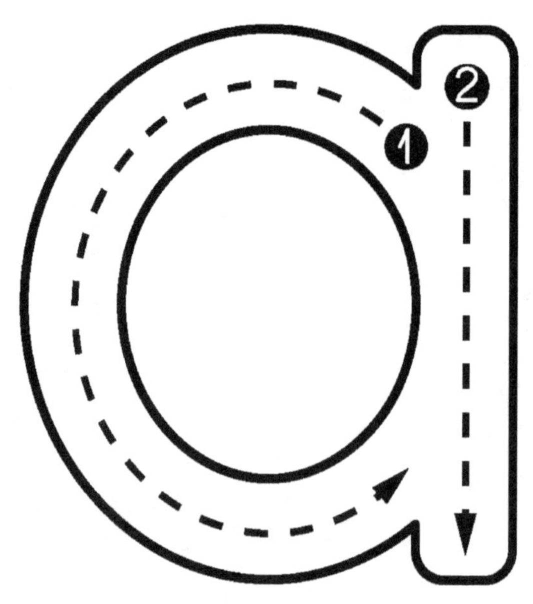

a is for
alligator

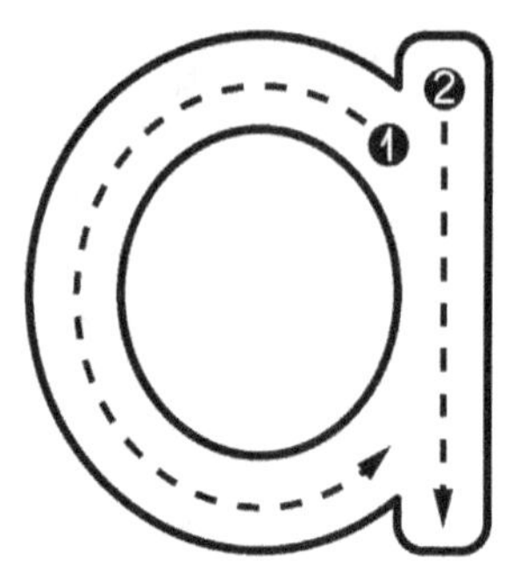

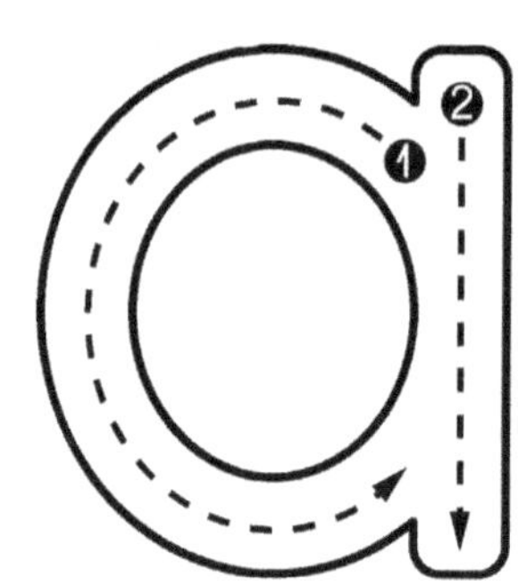

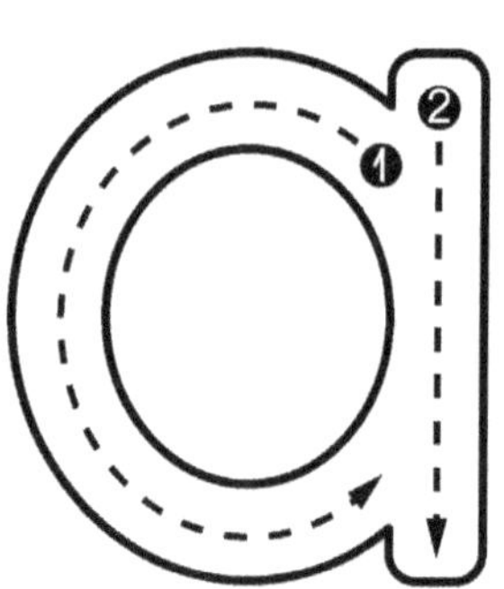

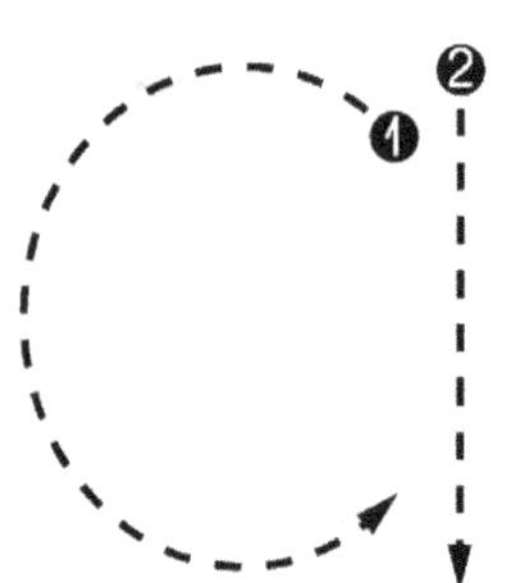

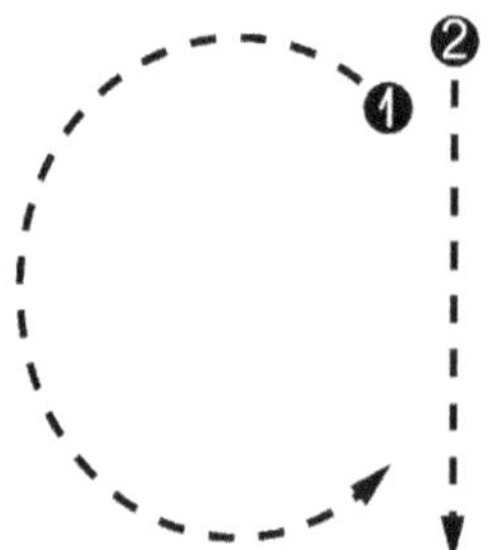

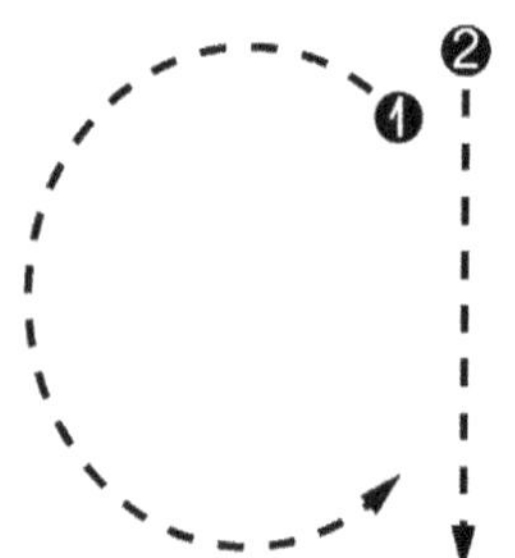

Excellent!
Now
you can write

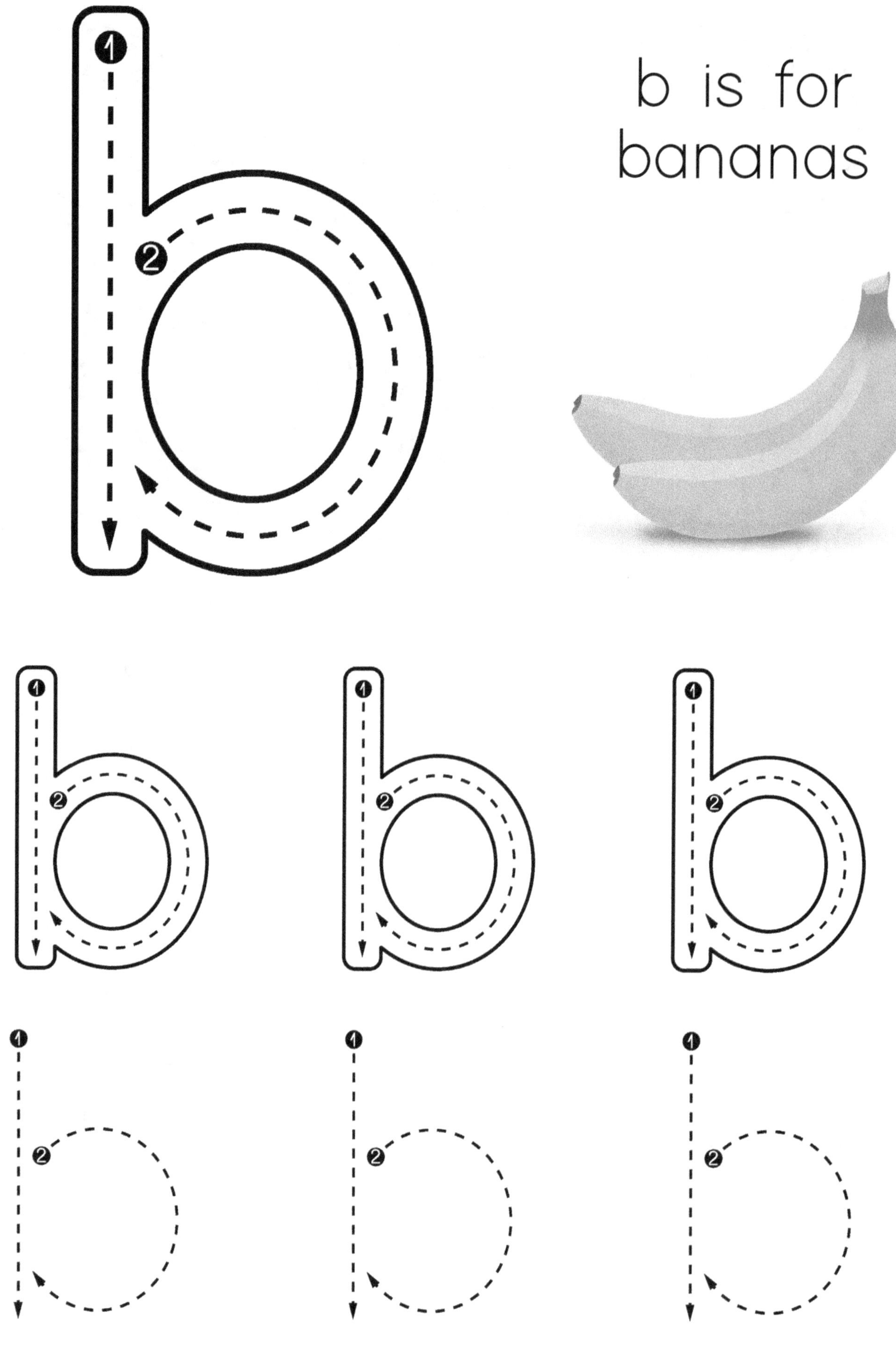
b is for
bananas

b b b

c is for
cat

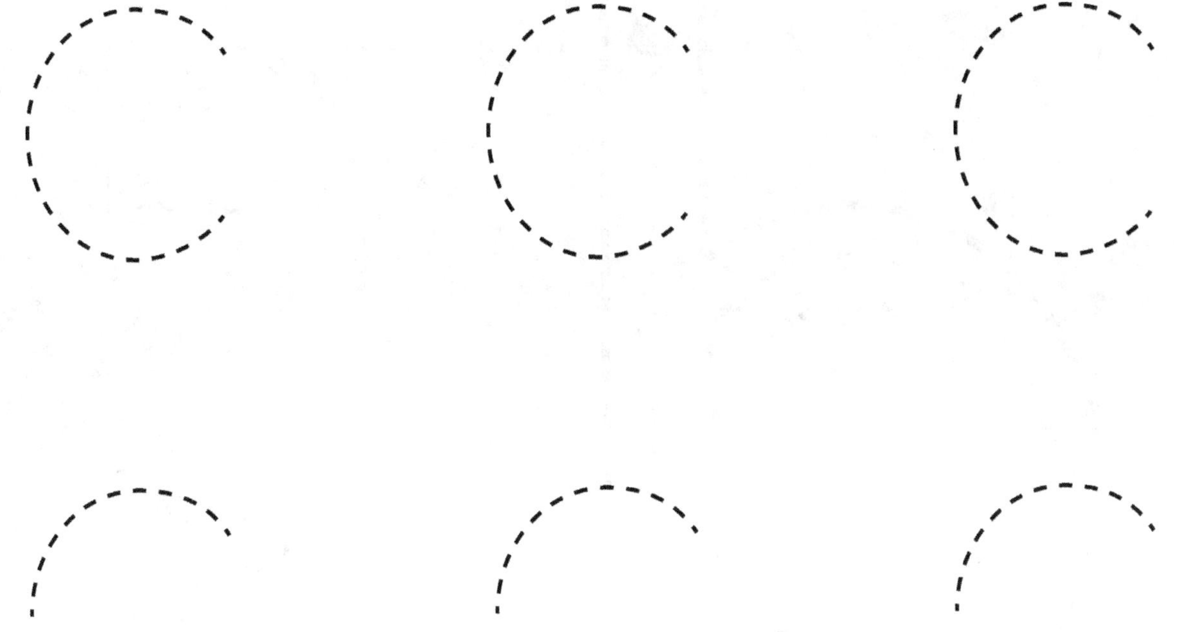

Excellent!
Now
you can write

c

d is for
dog

d d d

c c c

Excellent!
Now
you can write

d

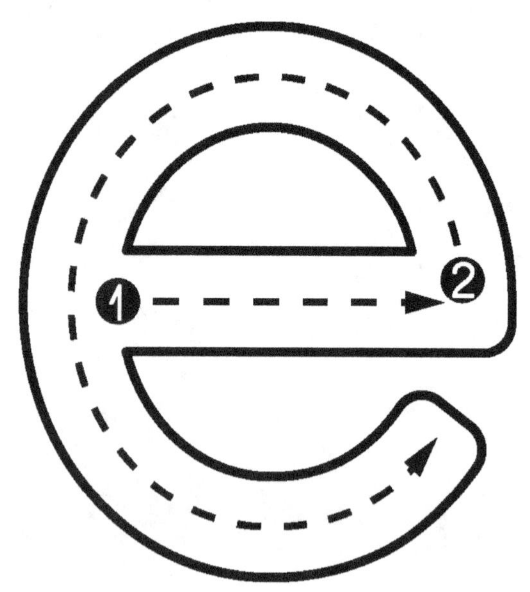

e is for
elephant

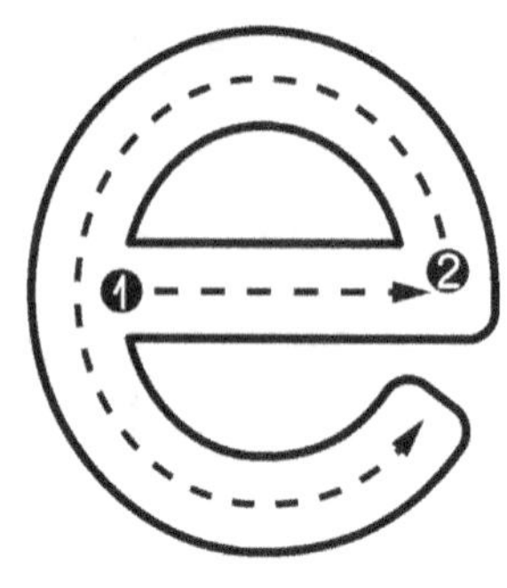

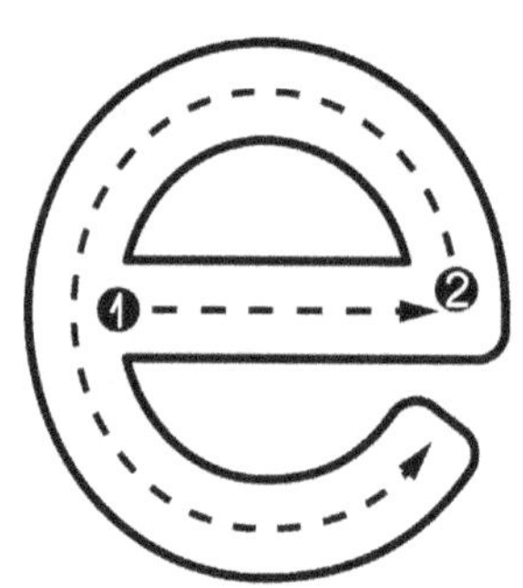

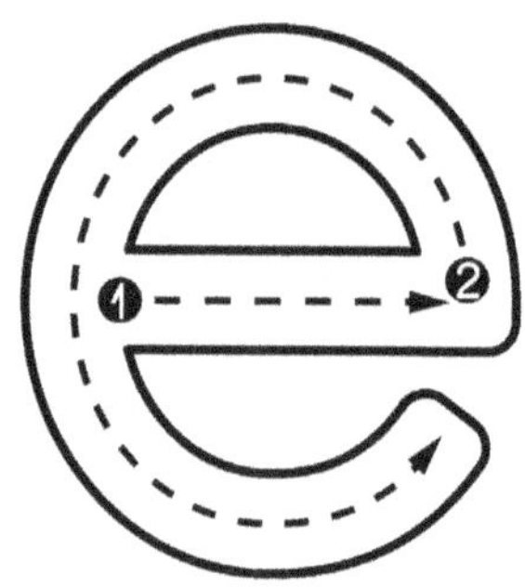

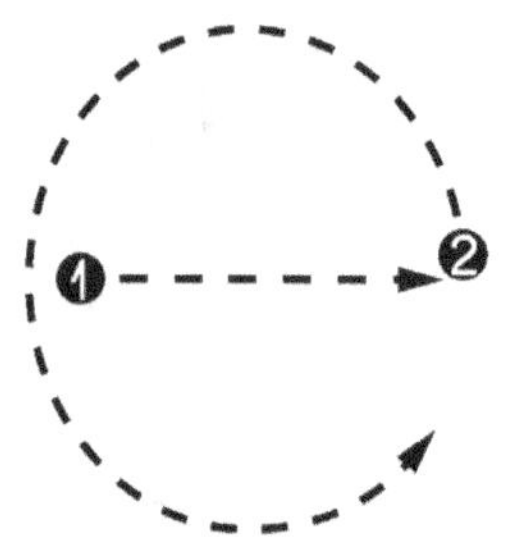

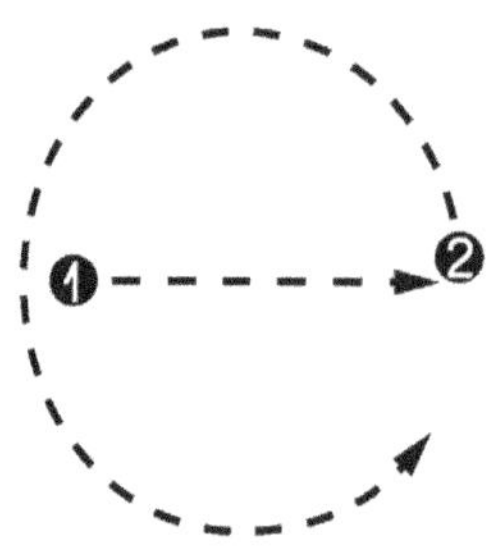

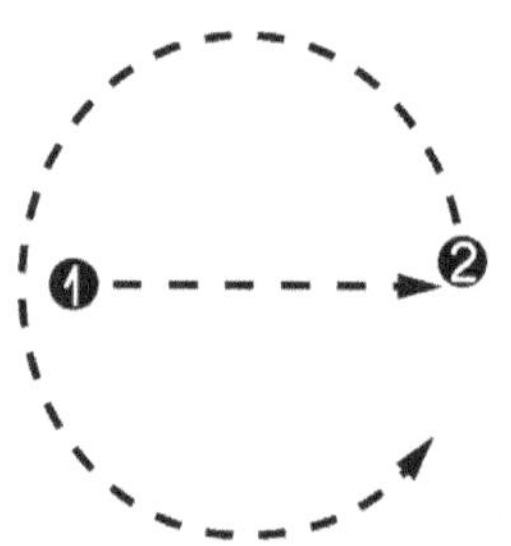

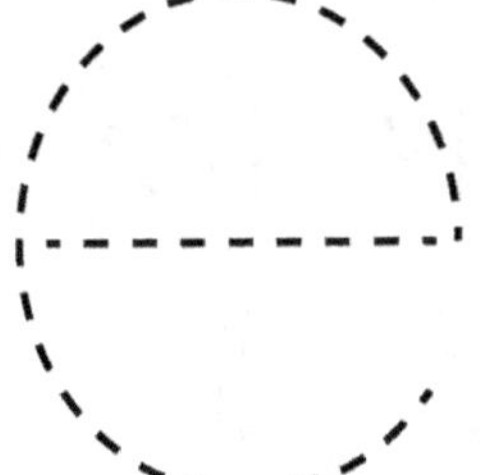 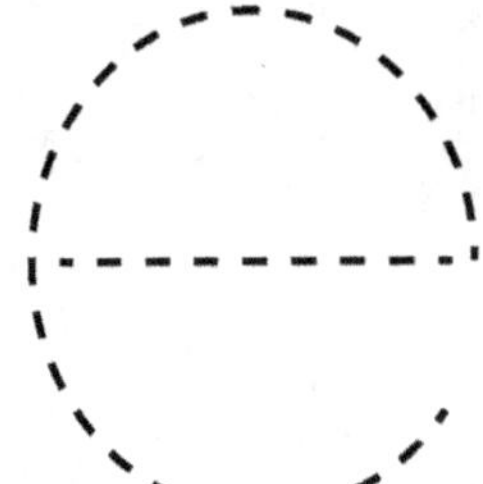 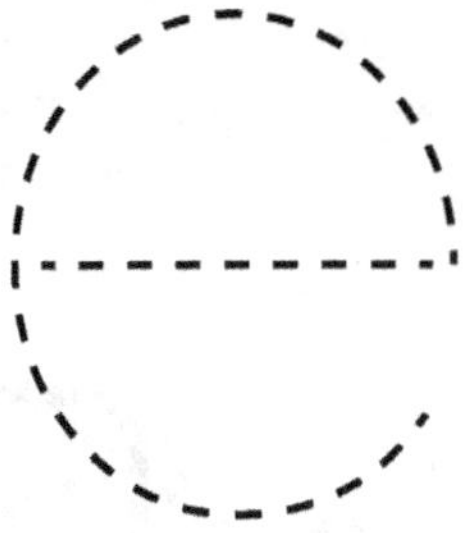

Excellent!
Now
you can write

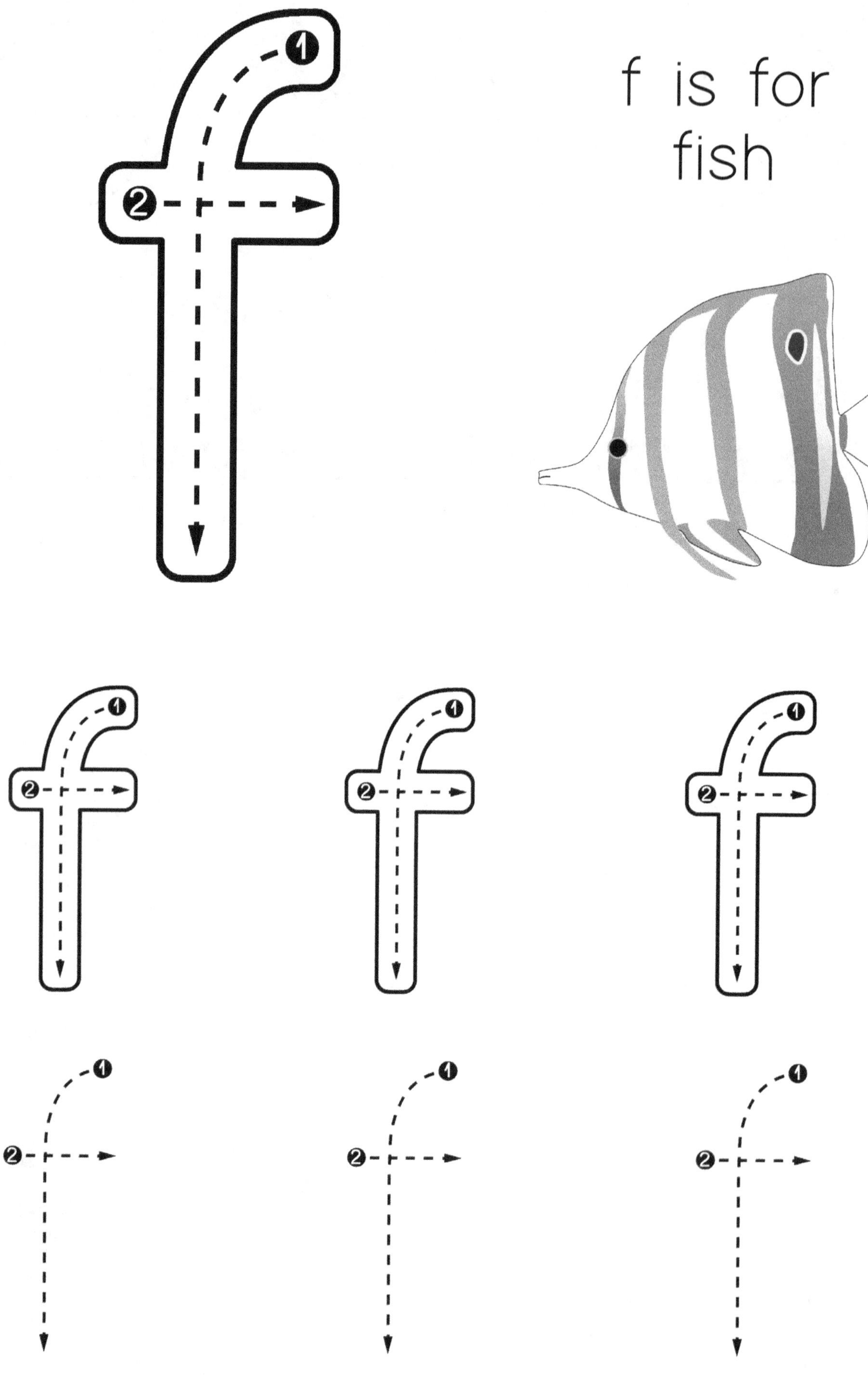

①
②
f is for
fish

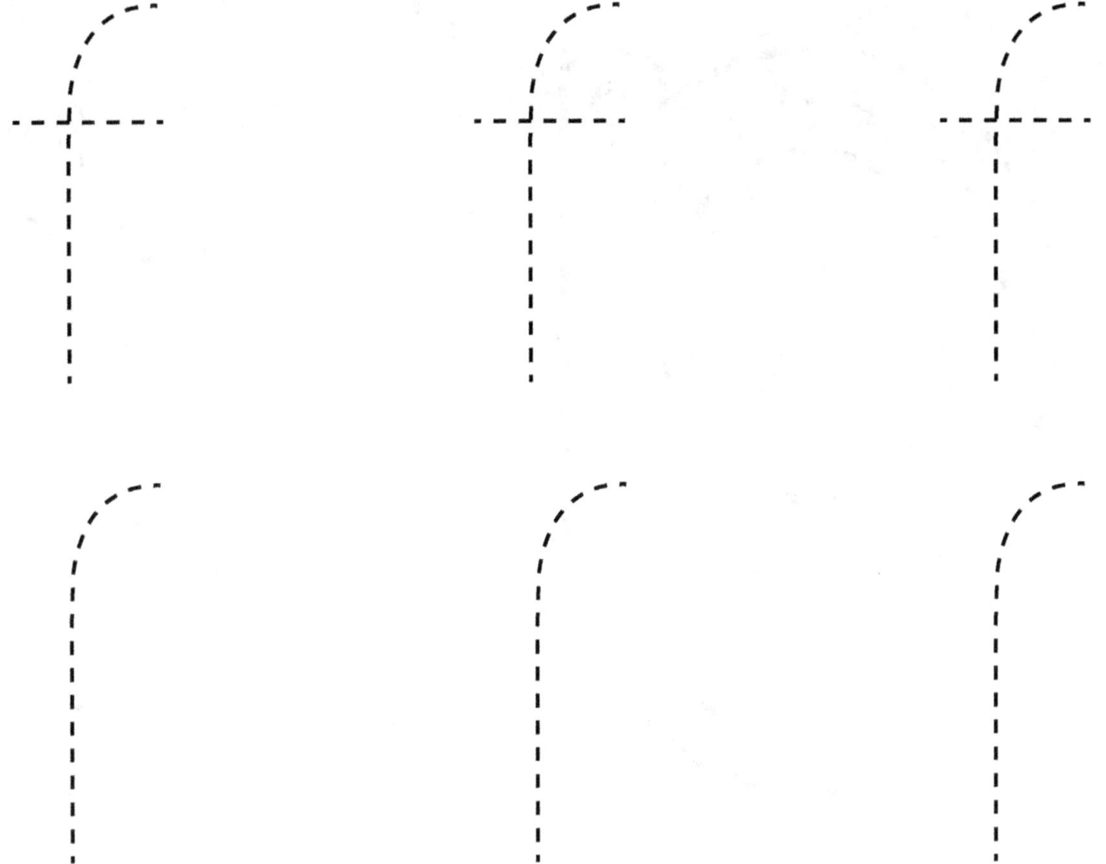

Excellent!
Now
you can write

f

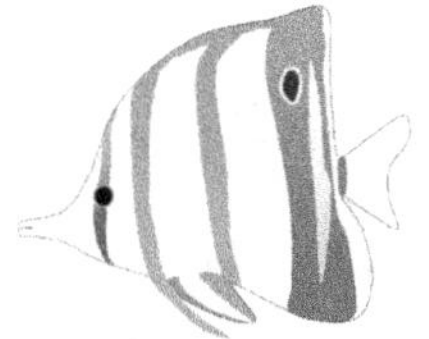

g is for
giraffe

Excellent!
Now
you can write

g

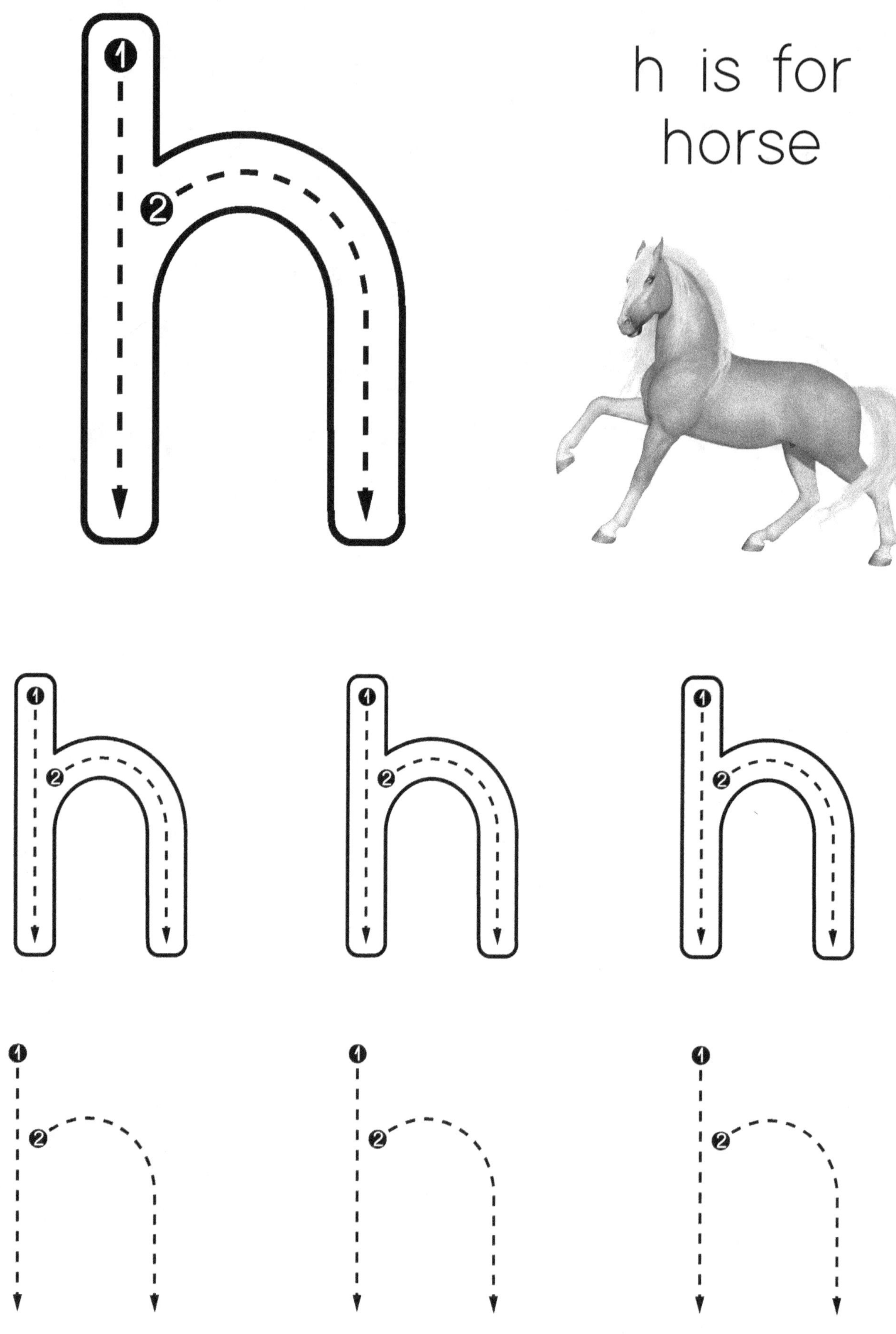
h is for
horse

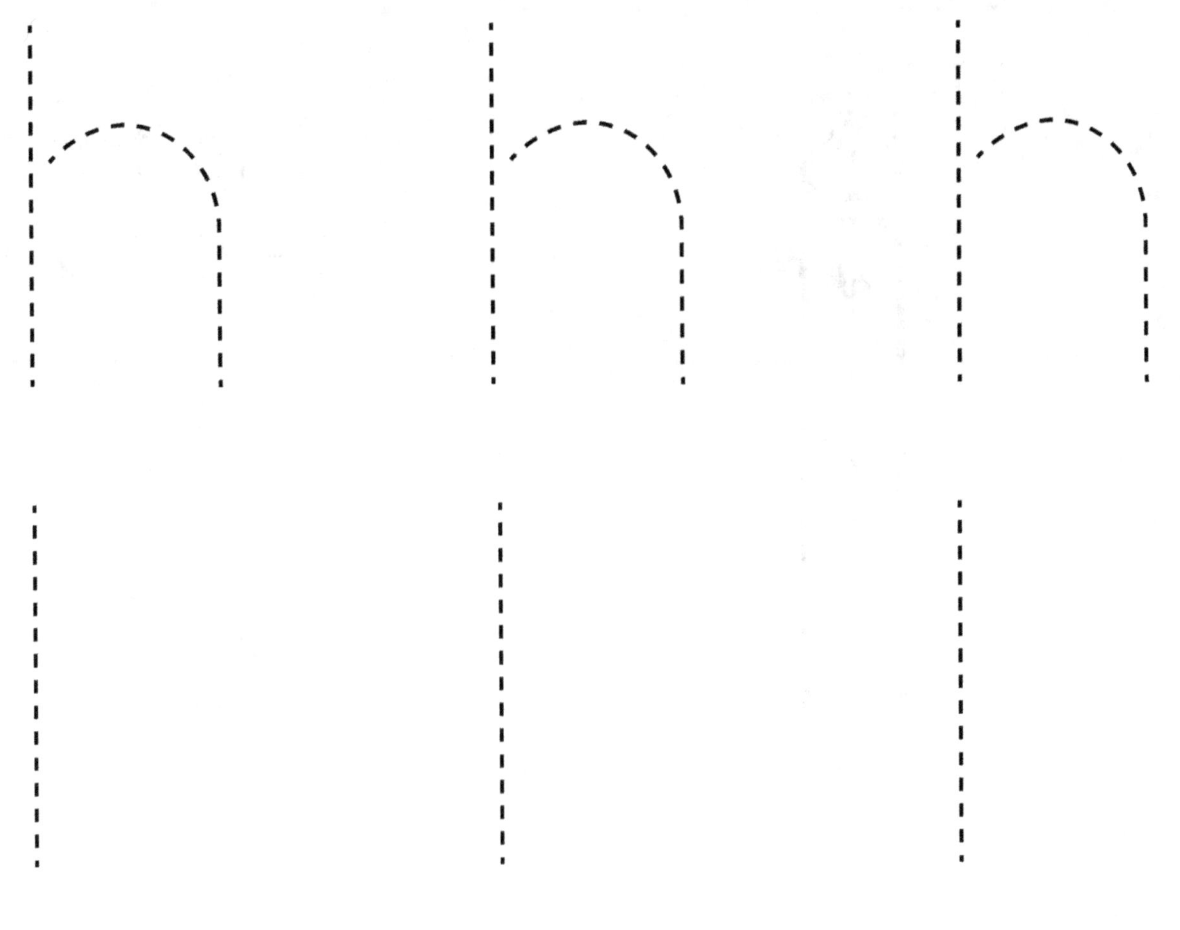

Excellent!
Now
you can write

h

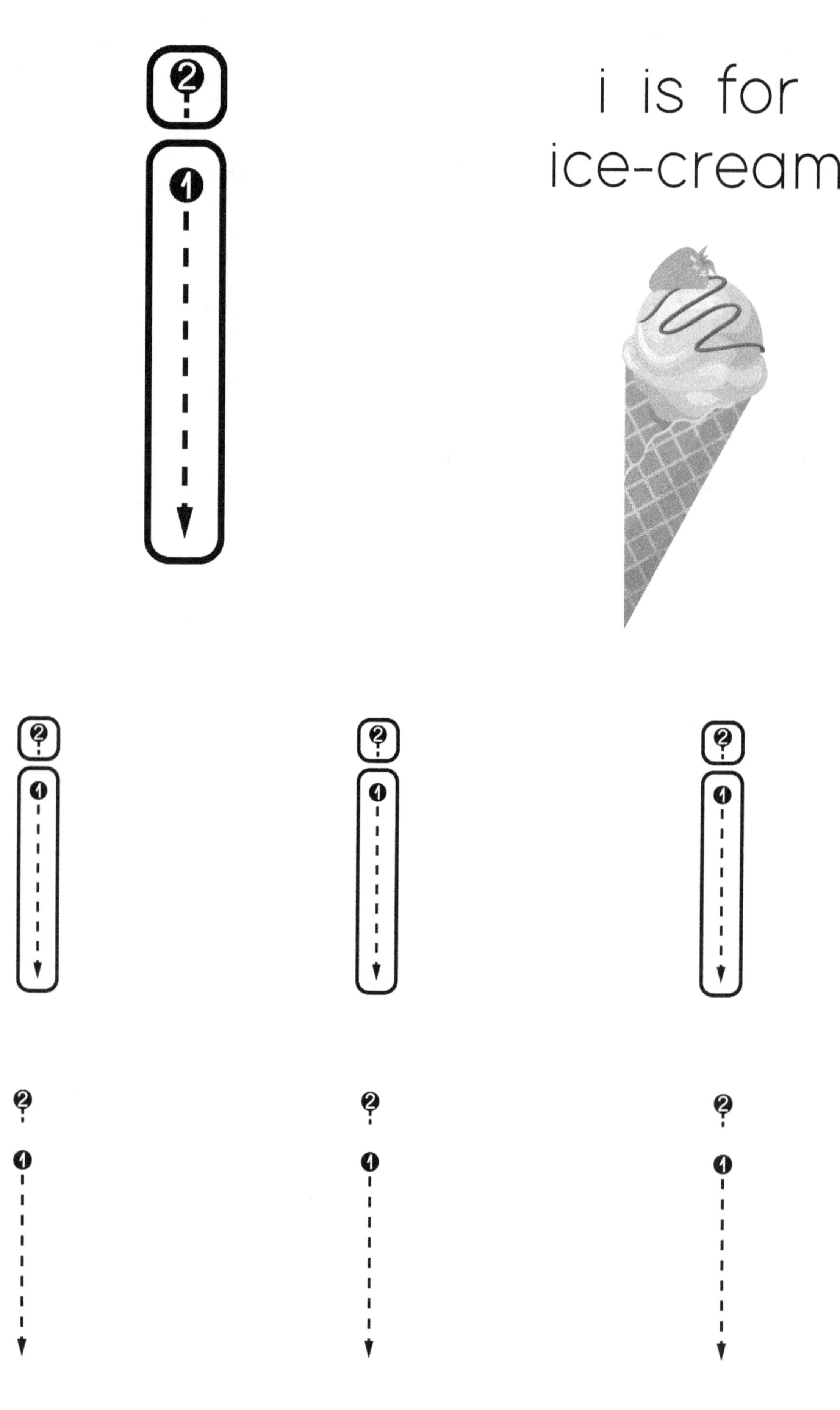

i is for
ice-cream

Excellent!
Now
you can write

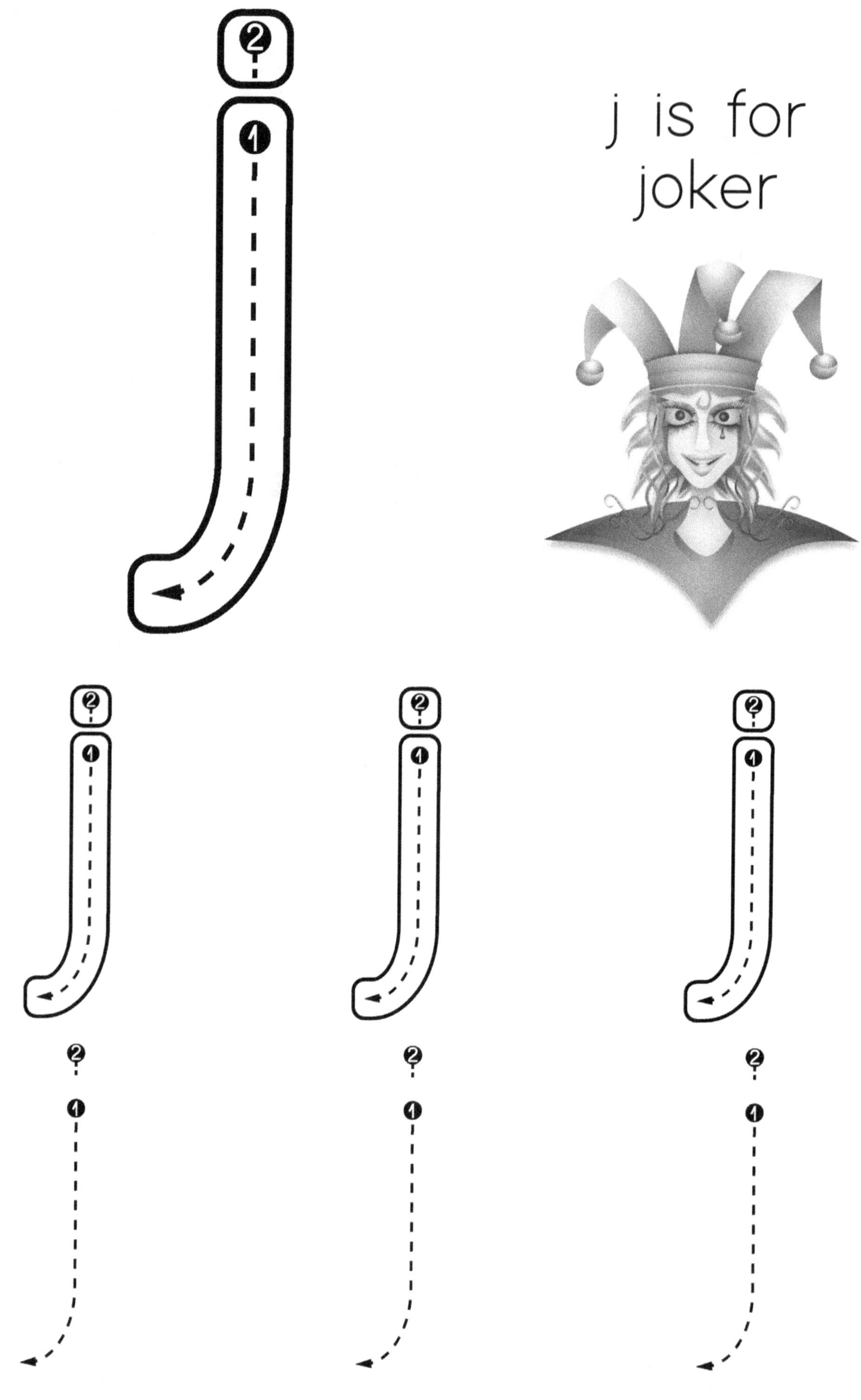
j is for
joker

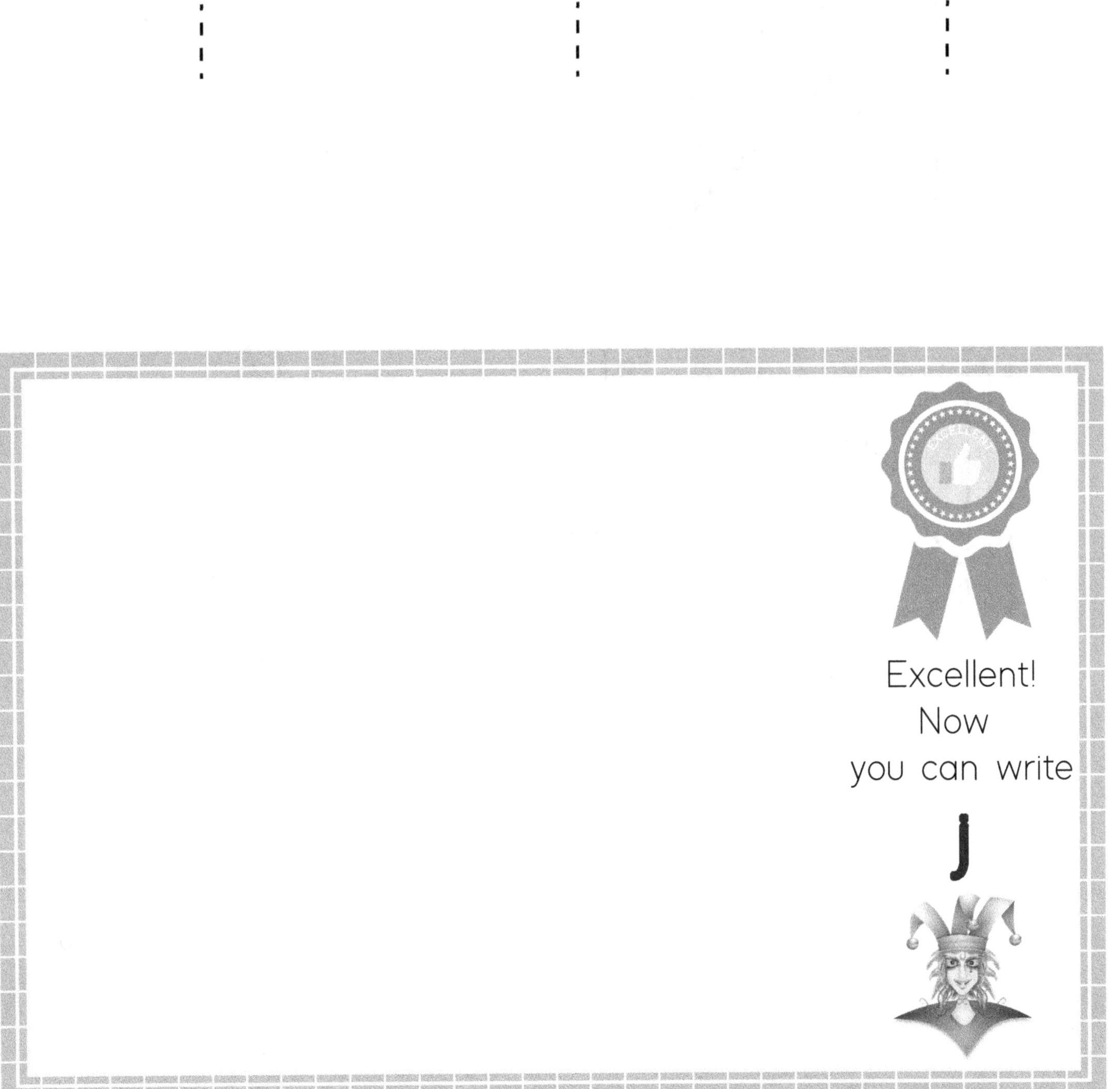

Excellent!
Now
you can write
J

k is for
kite

Excellent!
Now
you can write

k

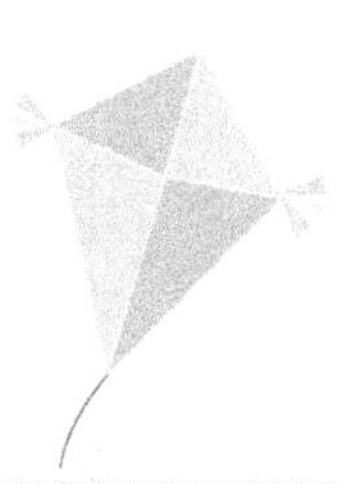

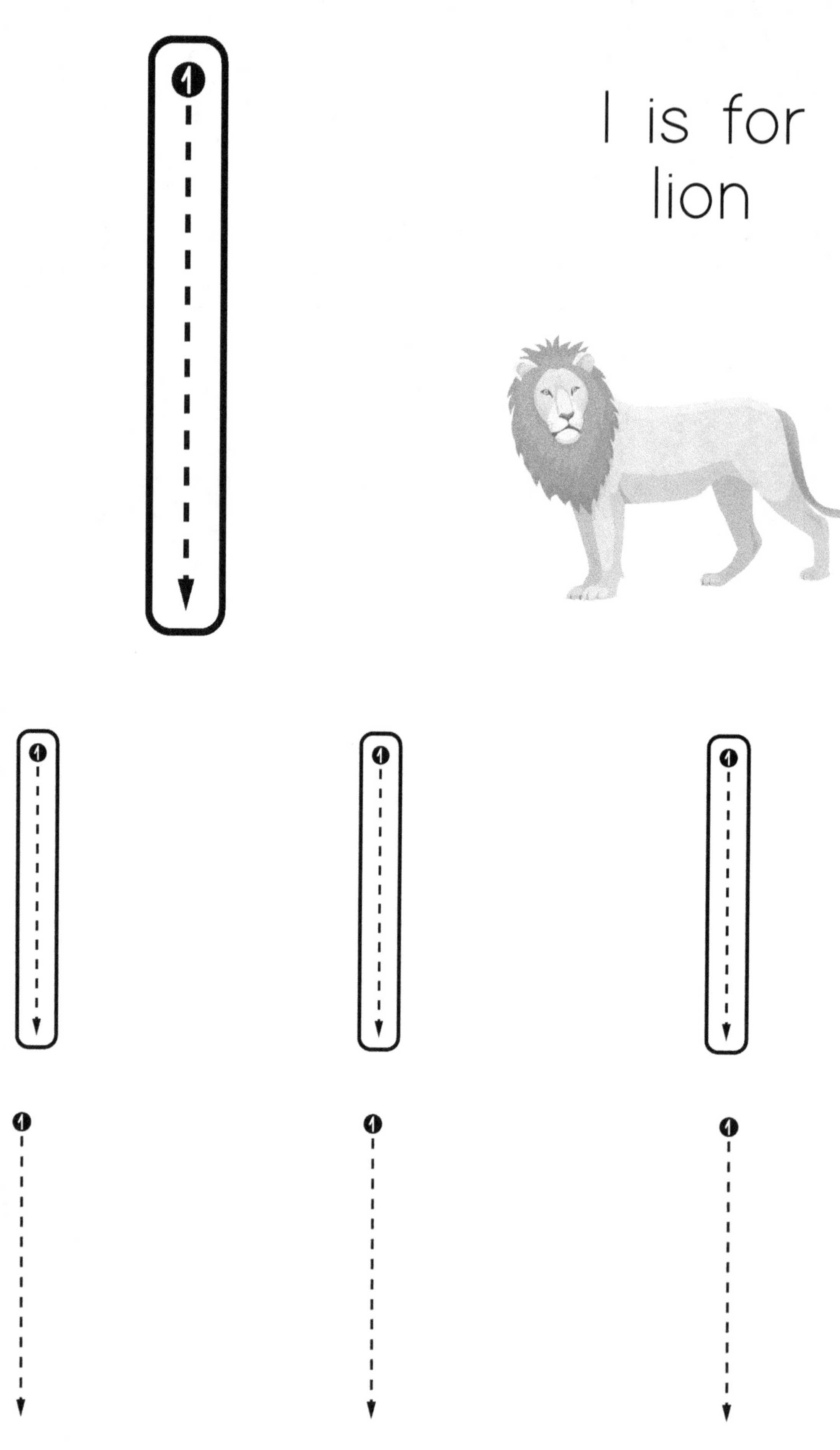

l is for
lion

Excellent!
Now
you can write
I

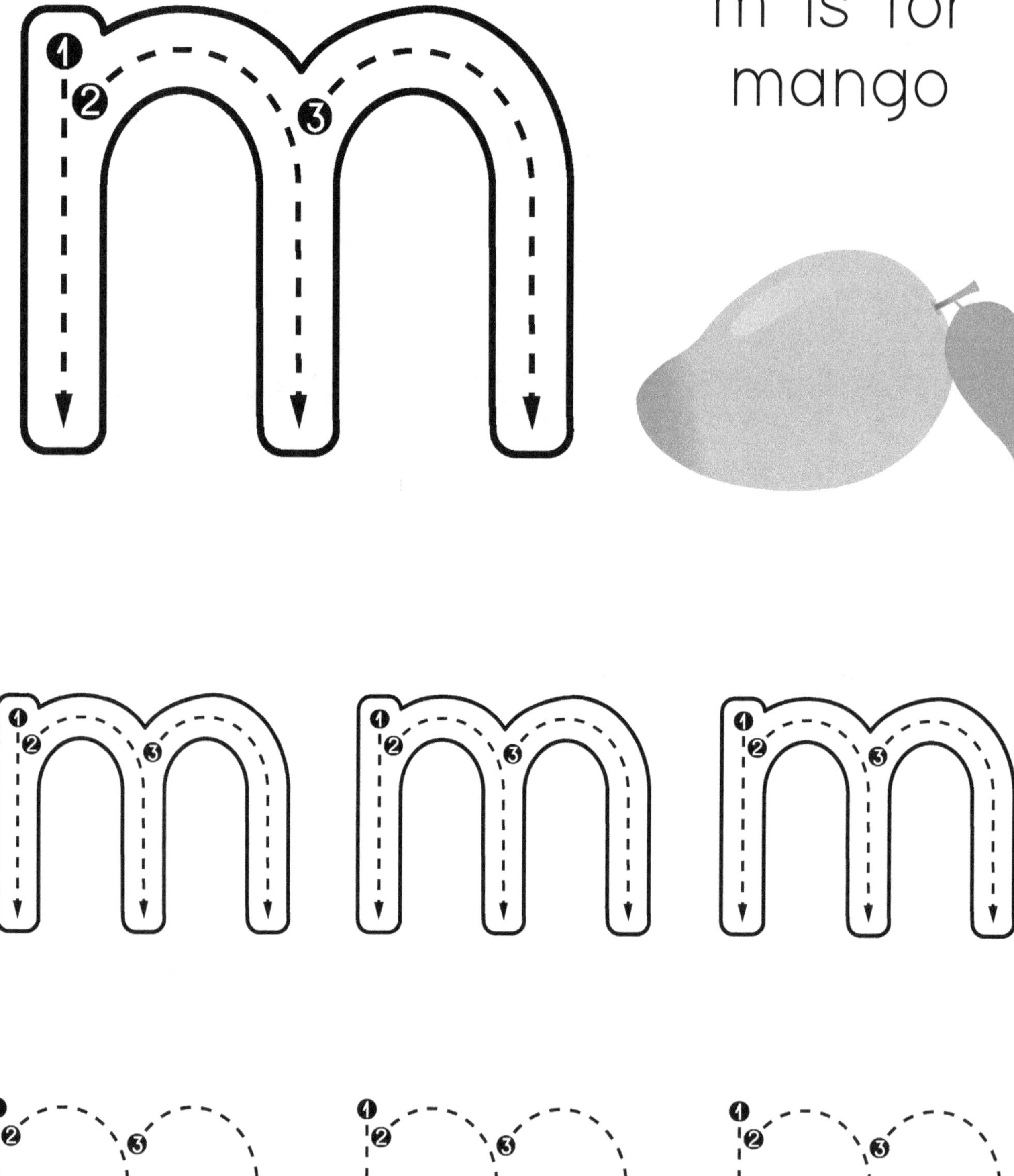

m is for
mango

Excellent!
Now
you can write

m

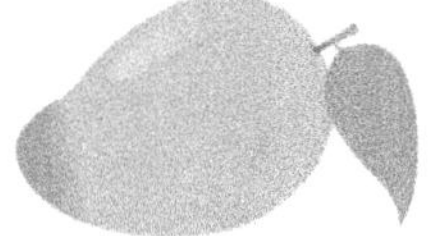

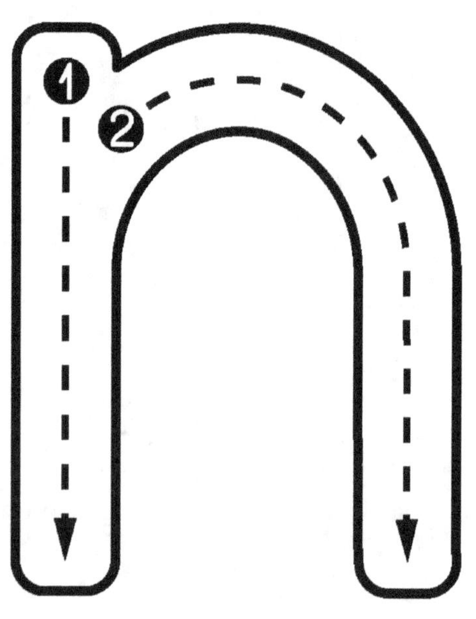

n is for
night

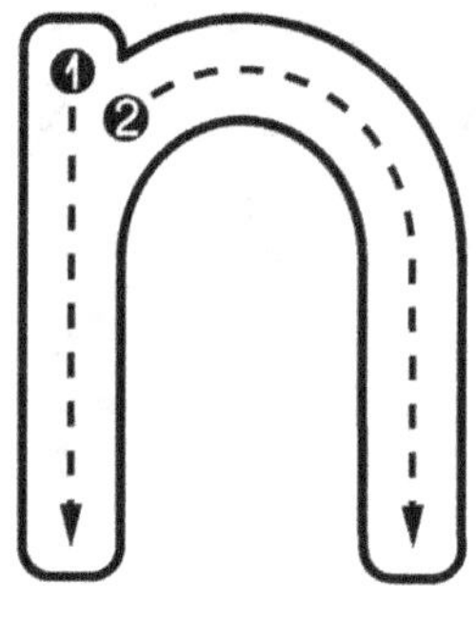

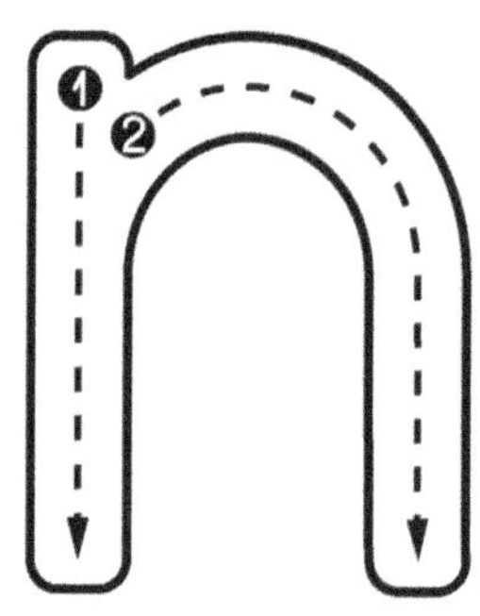

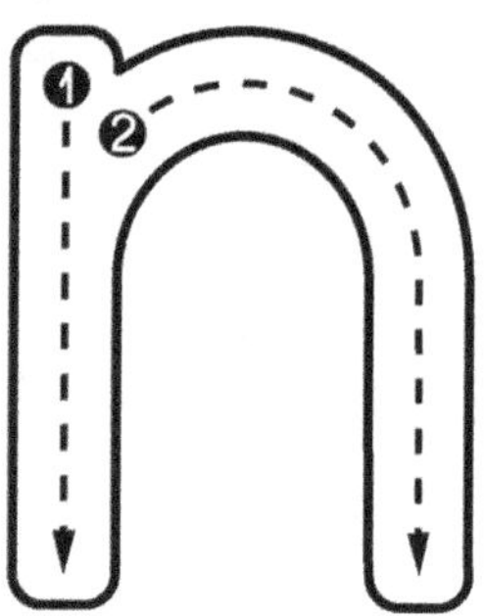

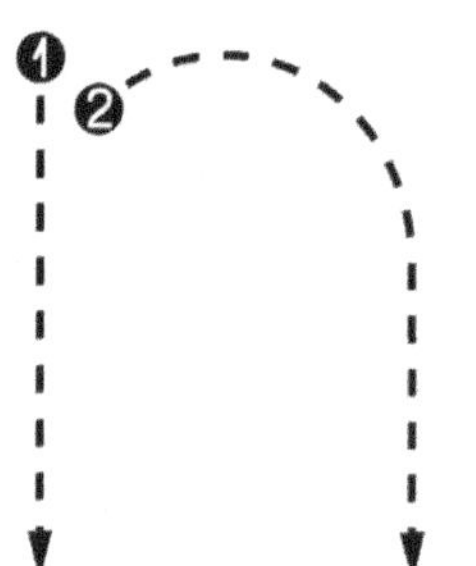

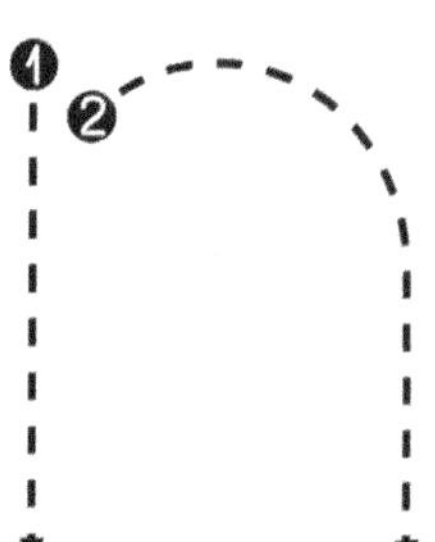

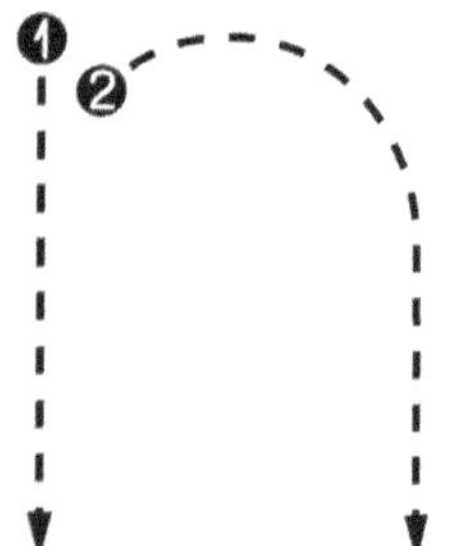

Excellent!
Now
you can write

n

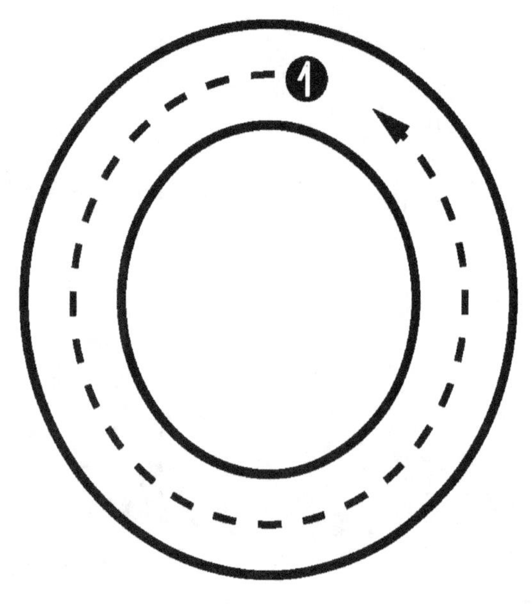

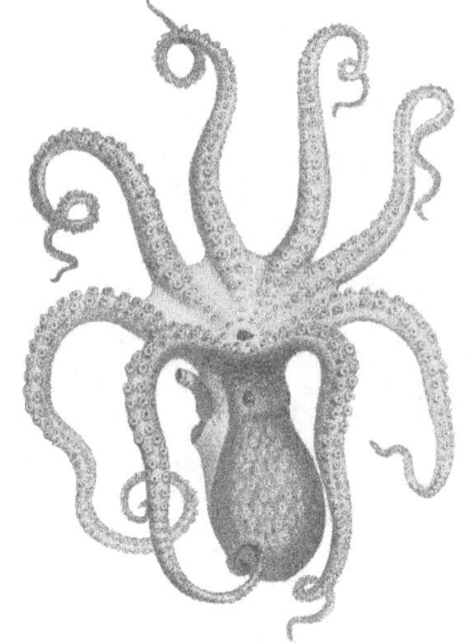

o is for
octopus

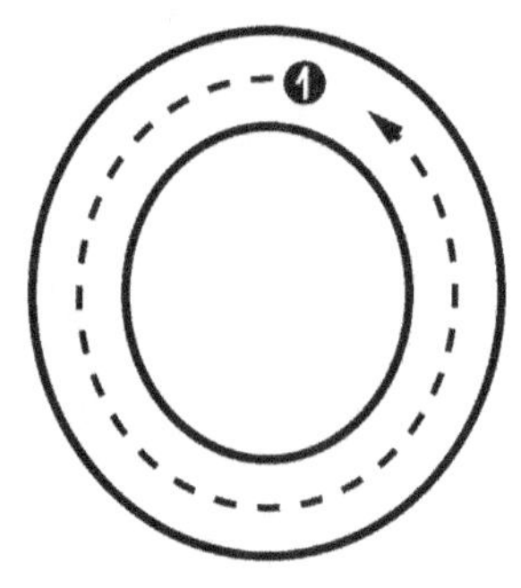 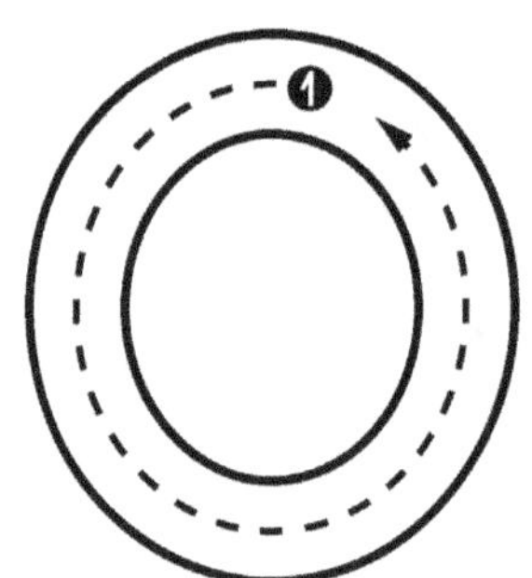 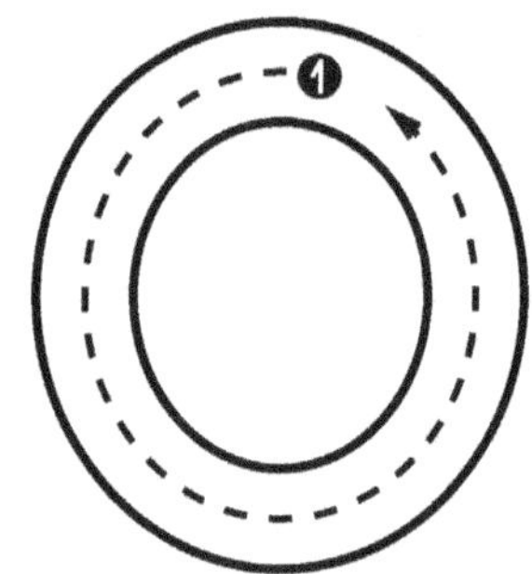

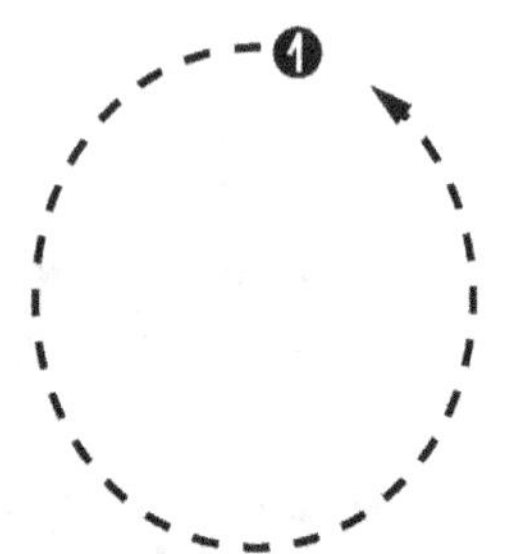 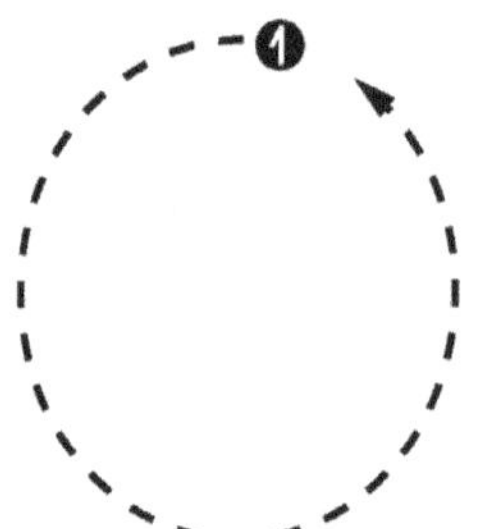 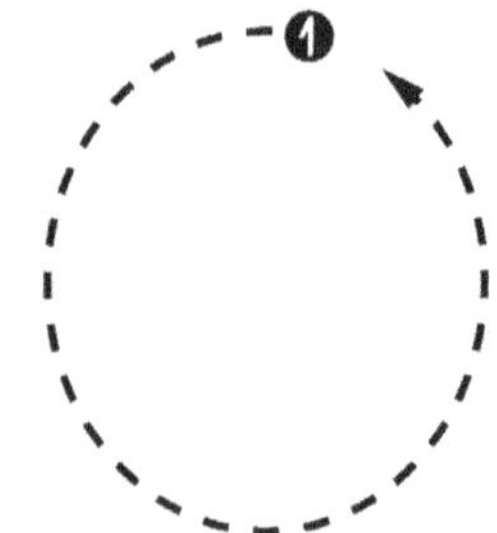

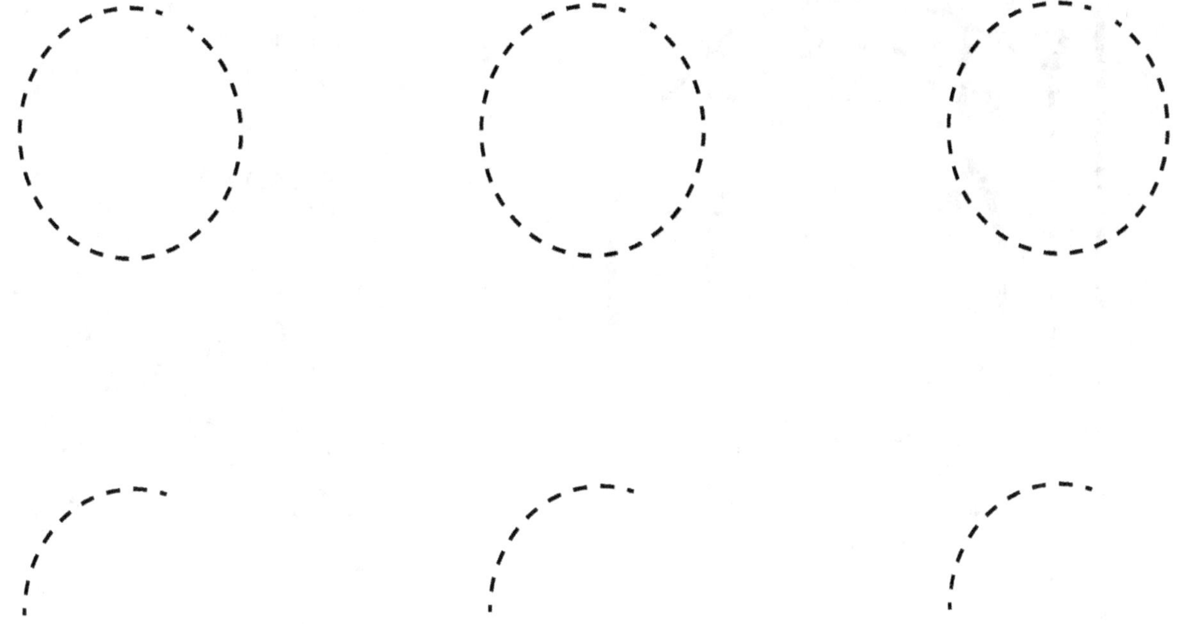

Excellent!
Now
you can write

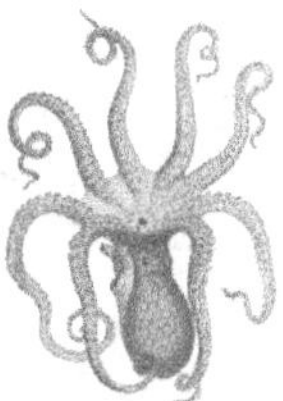

p is for
penguin

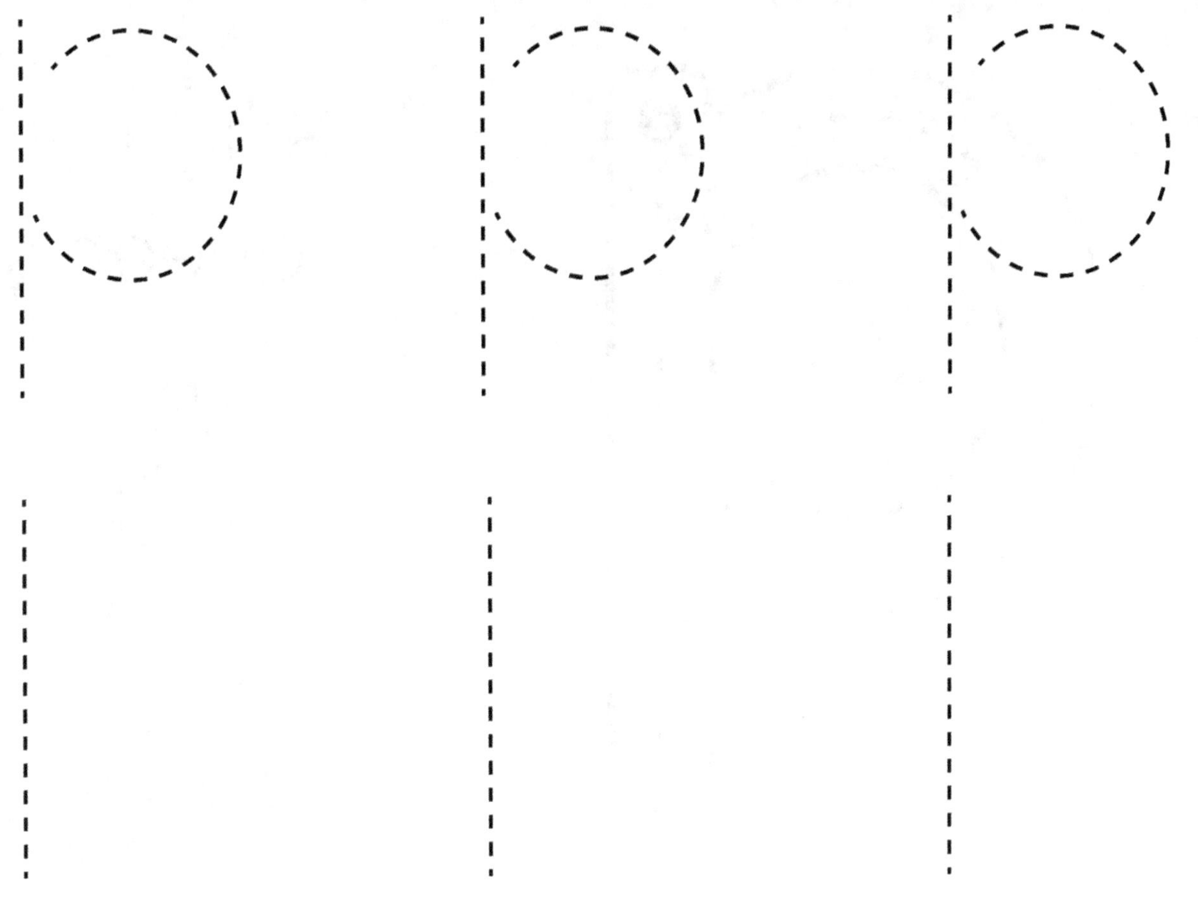

Excellent!
Now
you can write

p

q is for
queen

Excellent!
Now
you can write

q

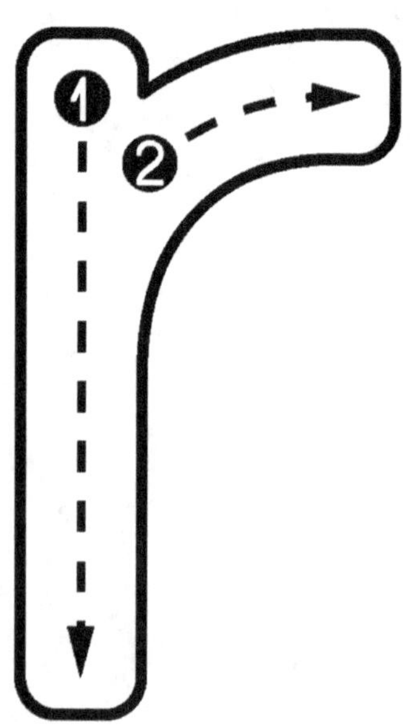

r is for
rat

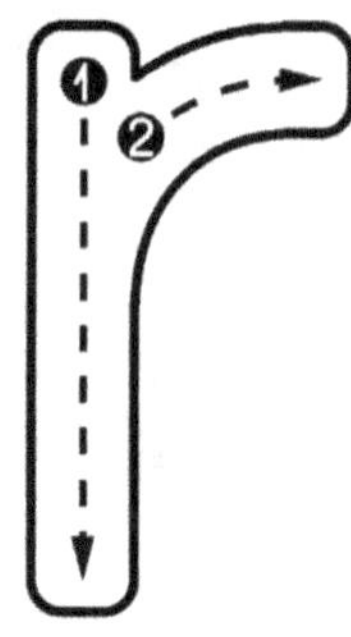

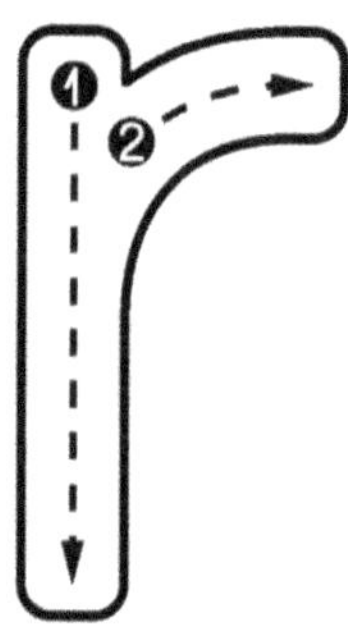

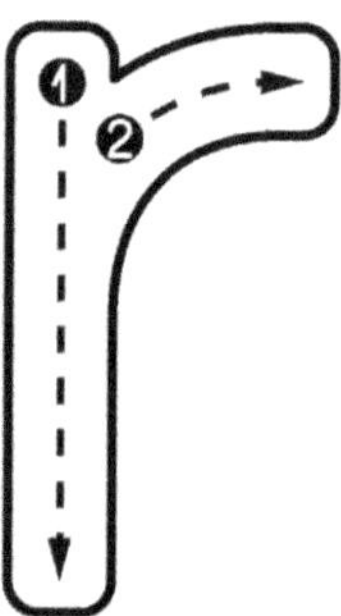

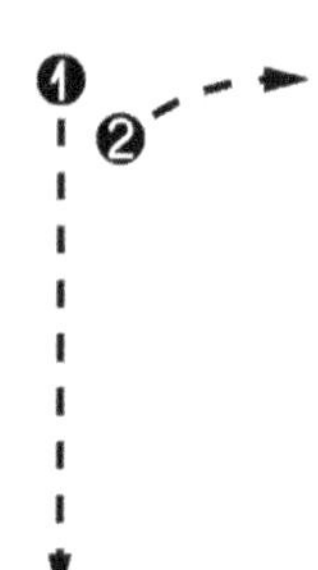

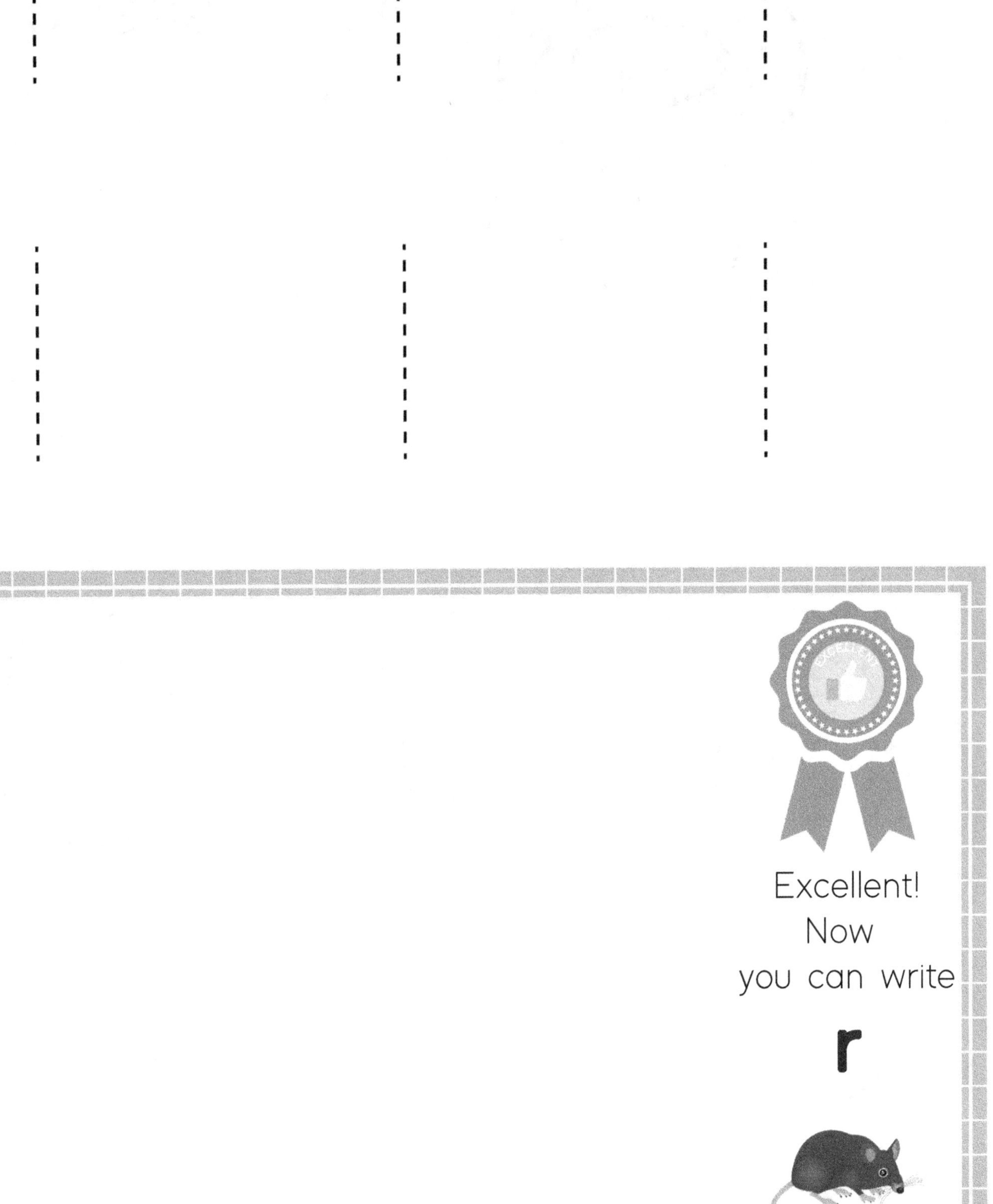

Excellent!
Now
you can write

r

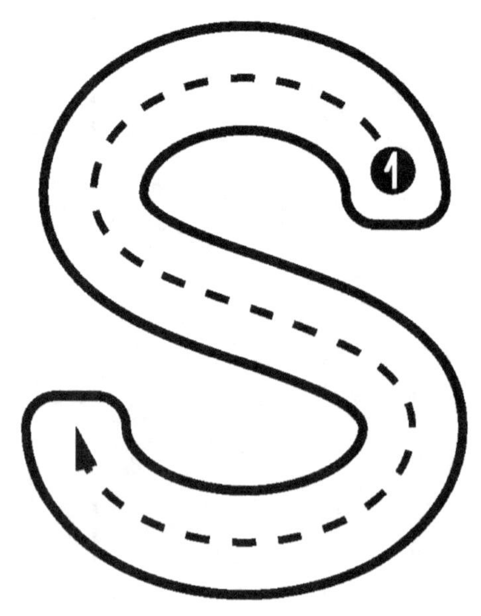

s is for
sheep

 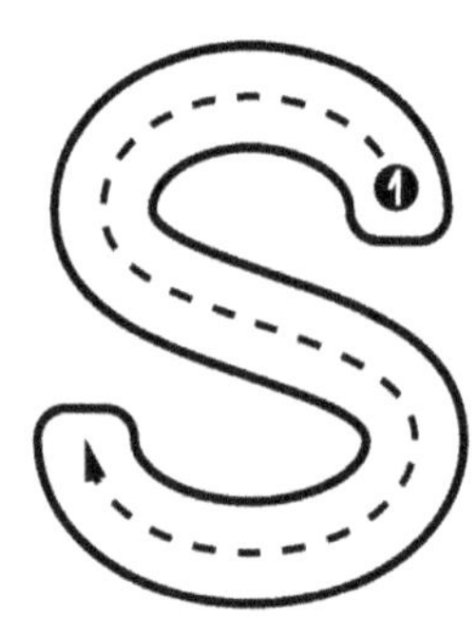 

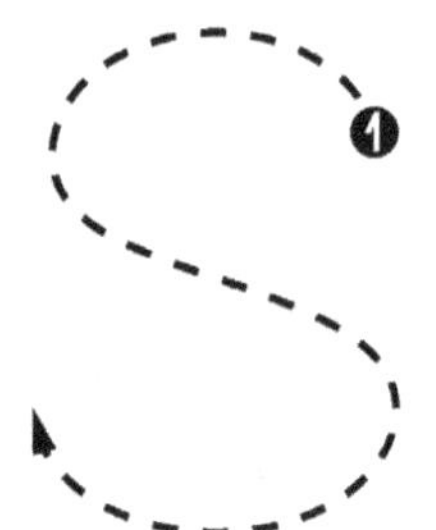 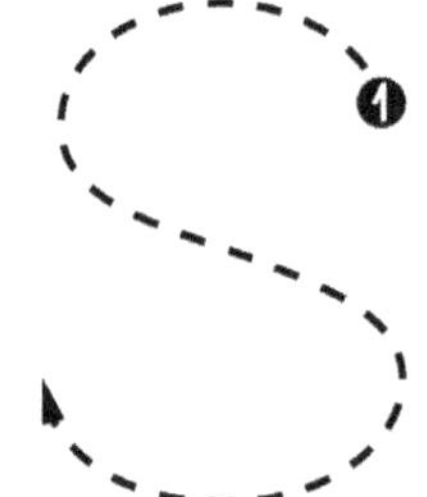 

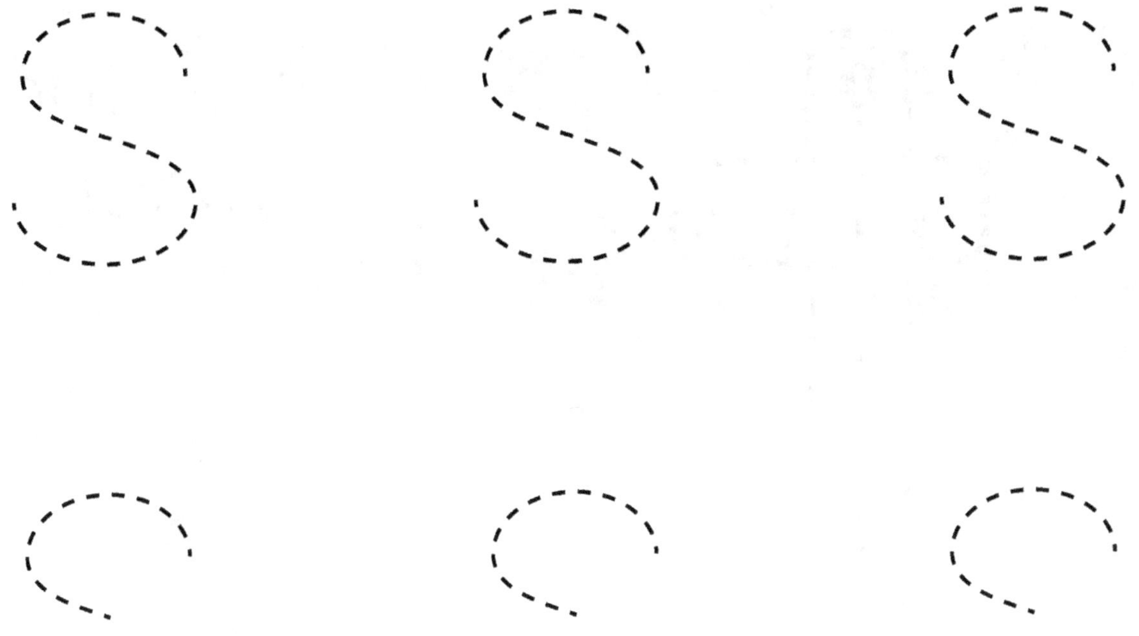

Excellent!
Now
you can write

s

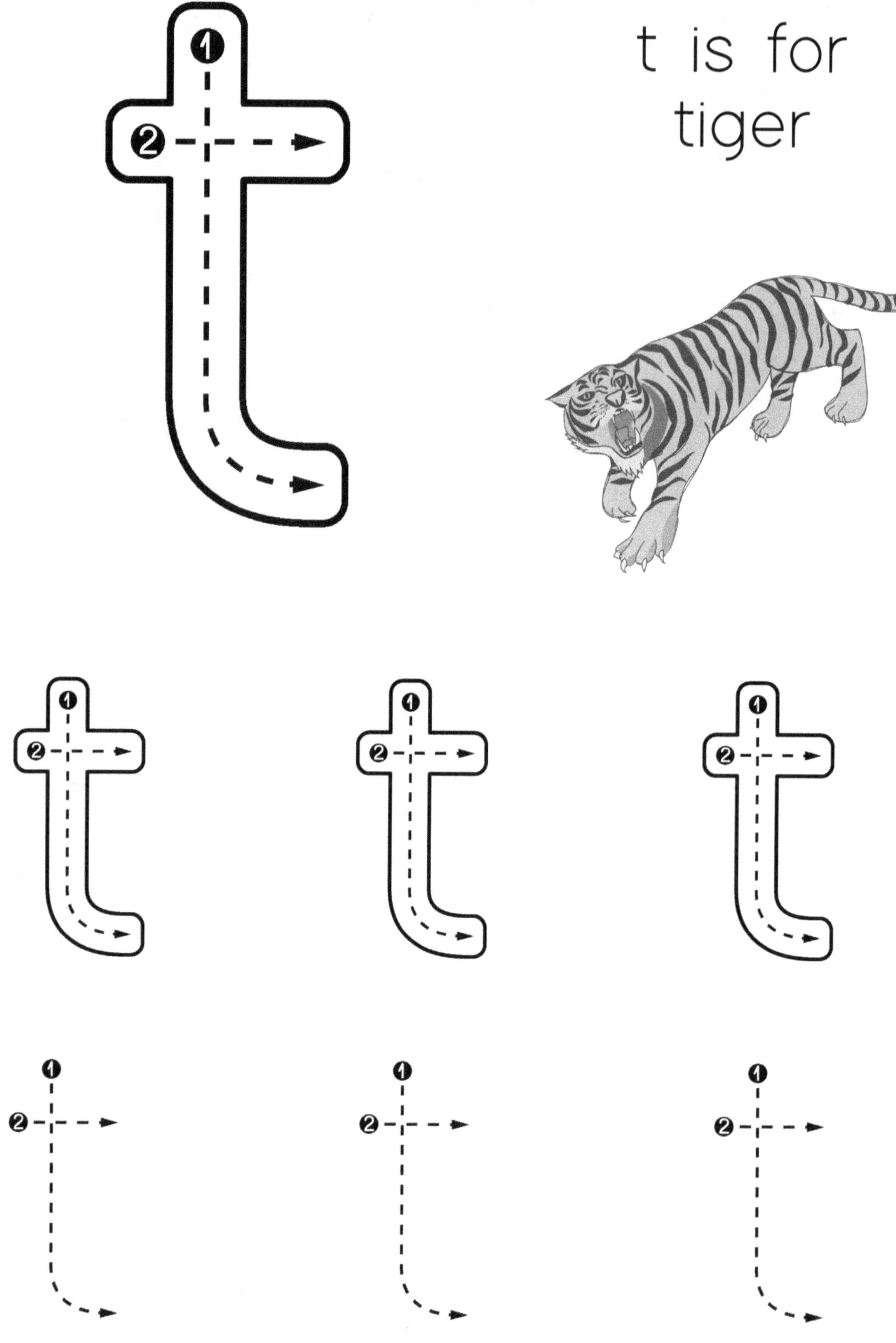

t is for
tiger

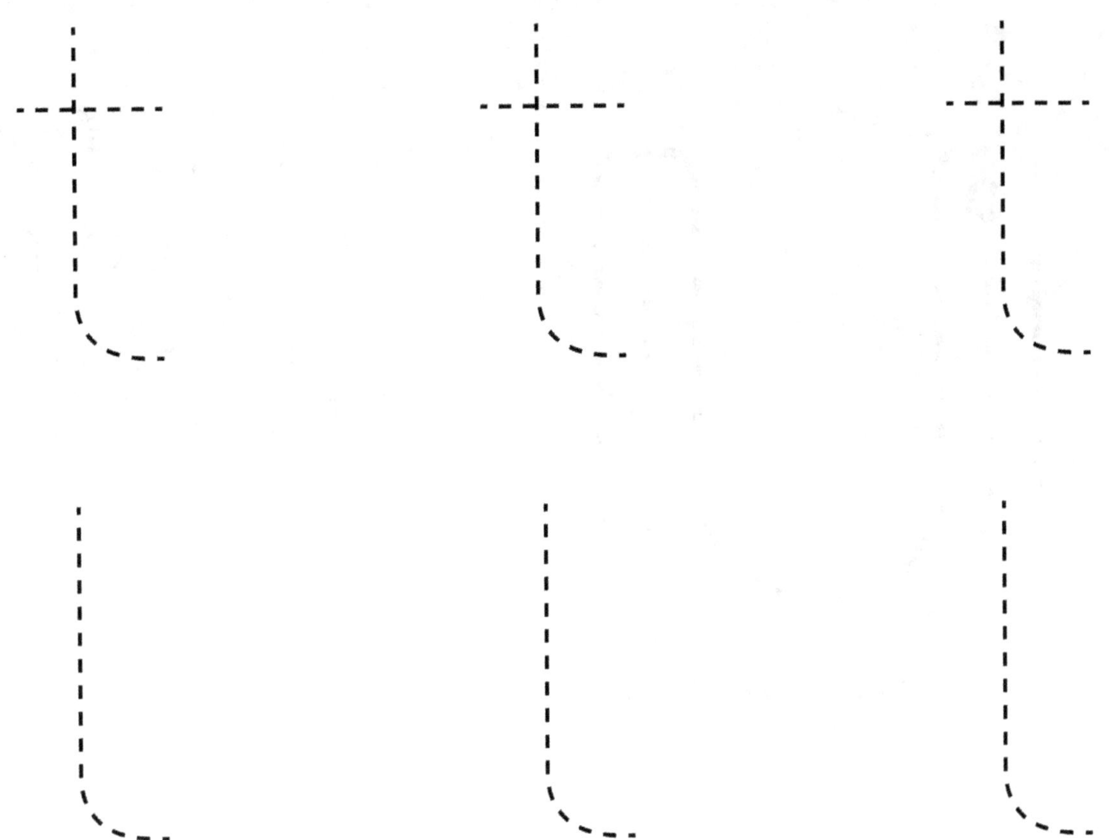

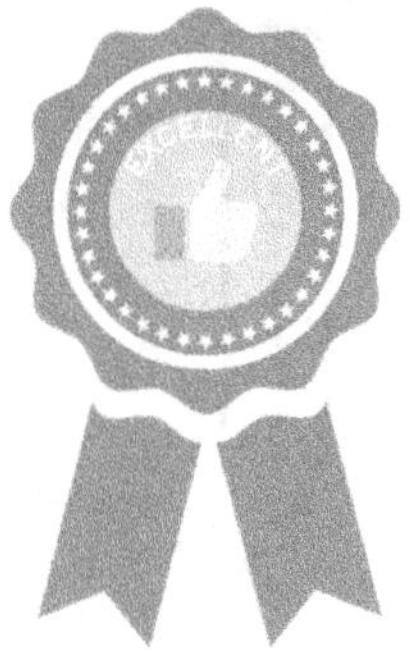

Excellent!
Now
you can write

t

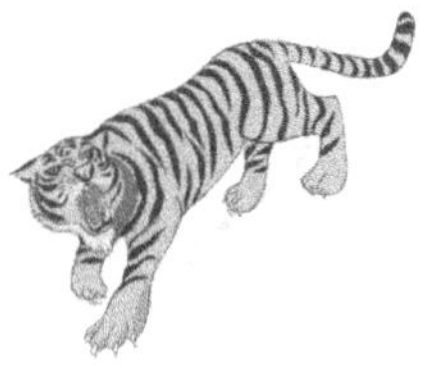

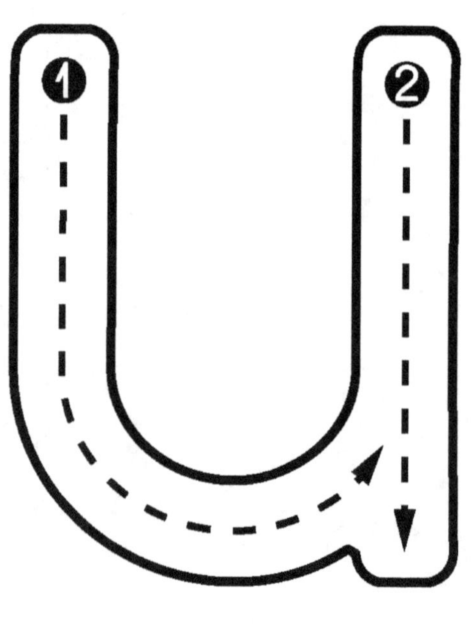

u is for
unicorn

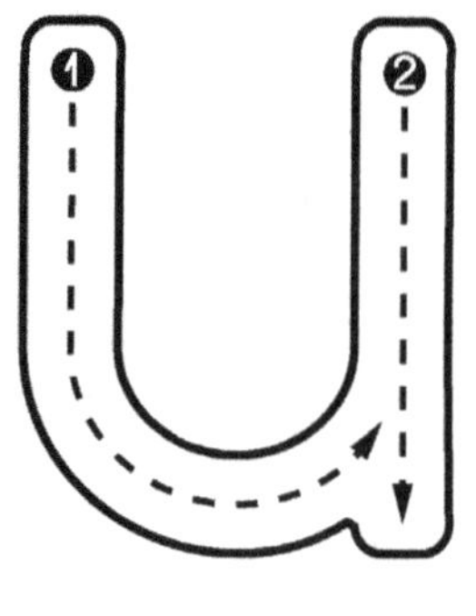

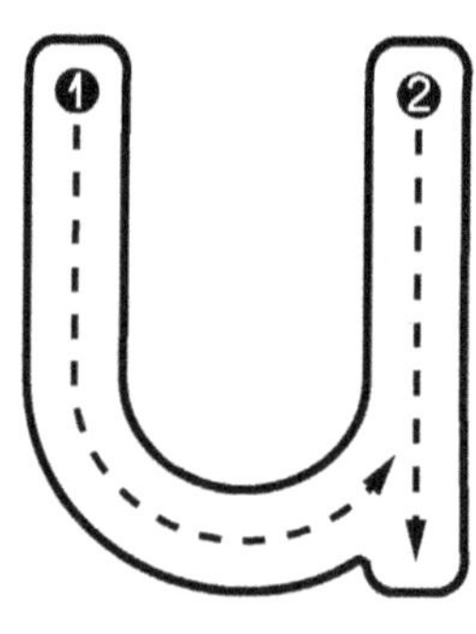

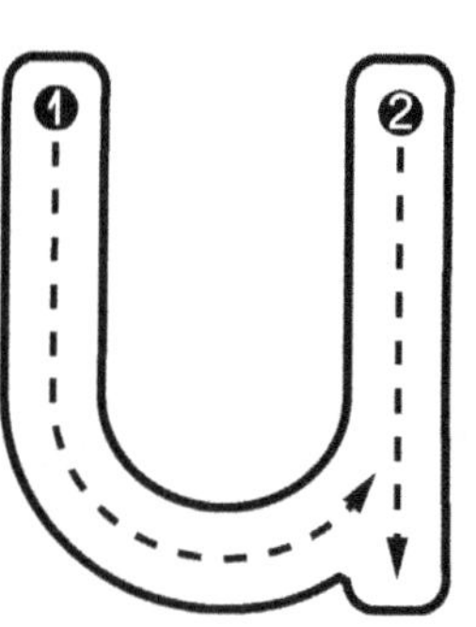

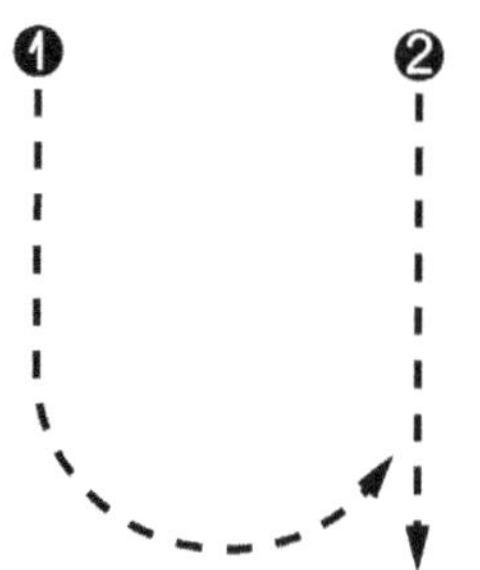

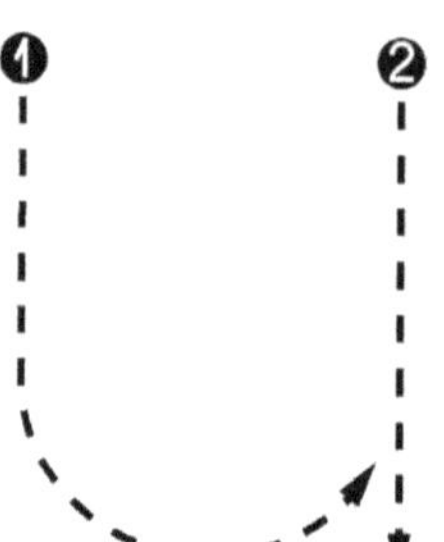

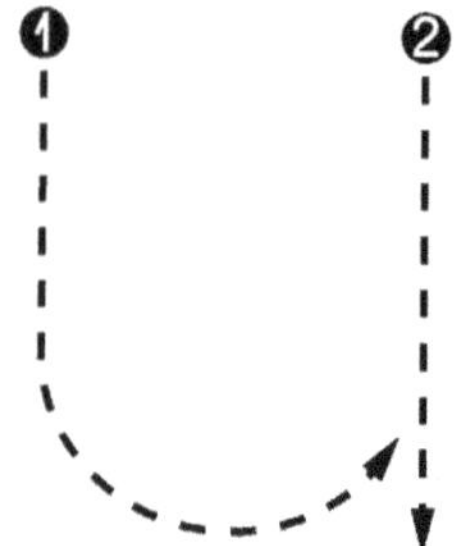

Excellent!
Now
you can write

U

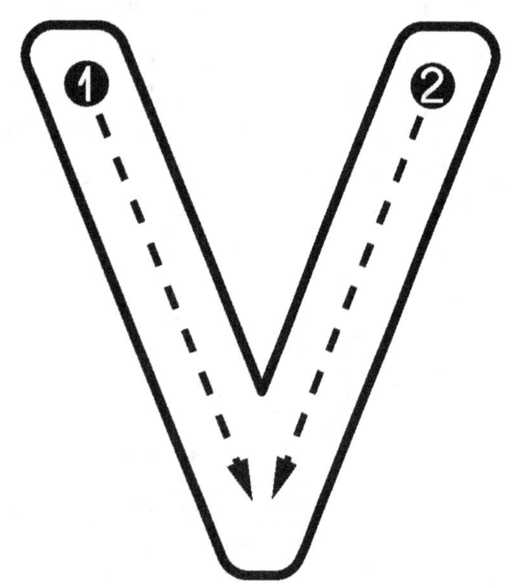

v is for
van

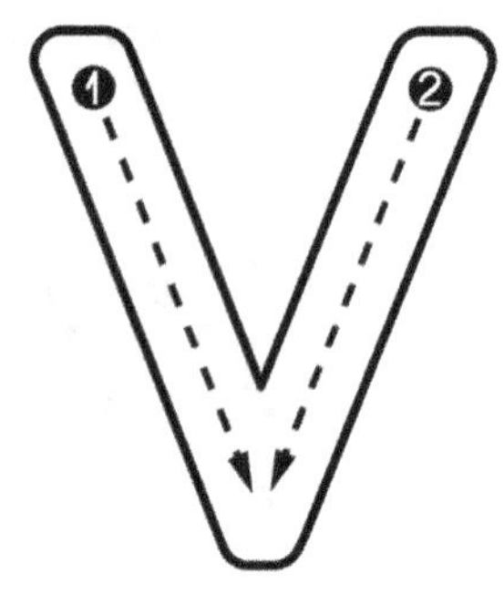
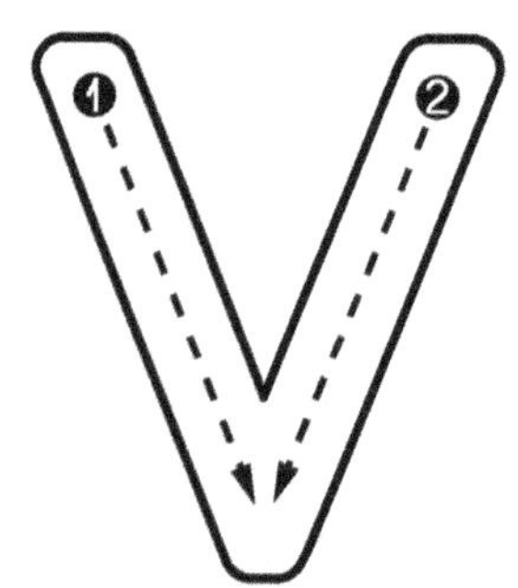
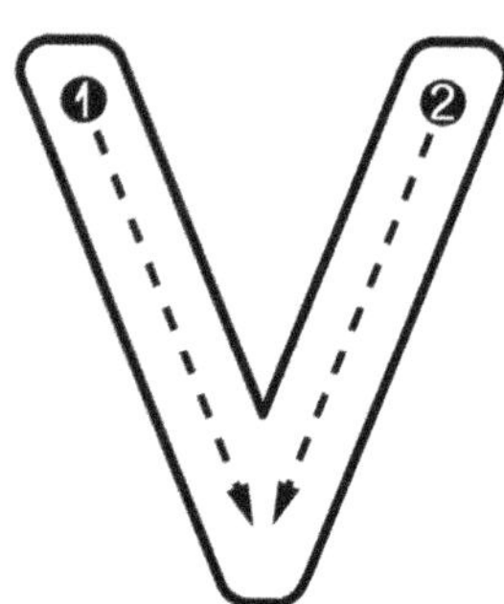
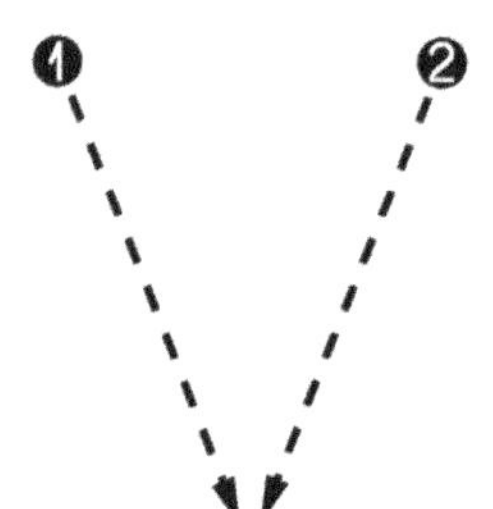
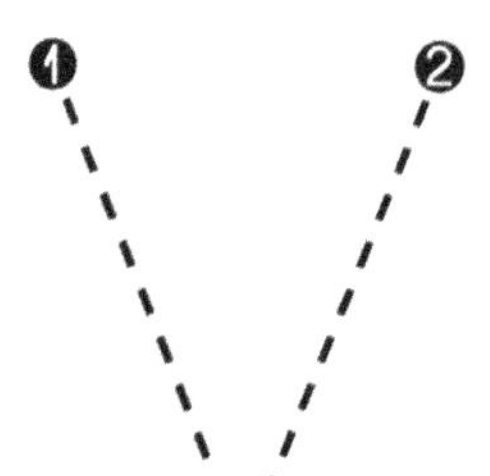
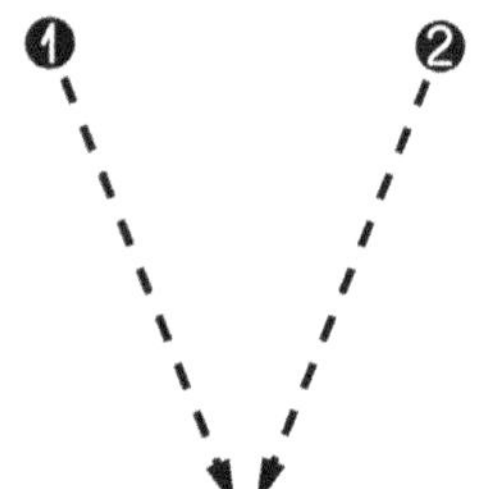

Excellent!
Now
you can write

V

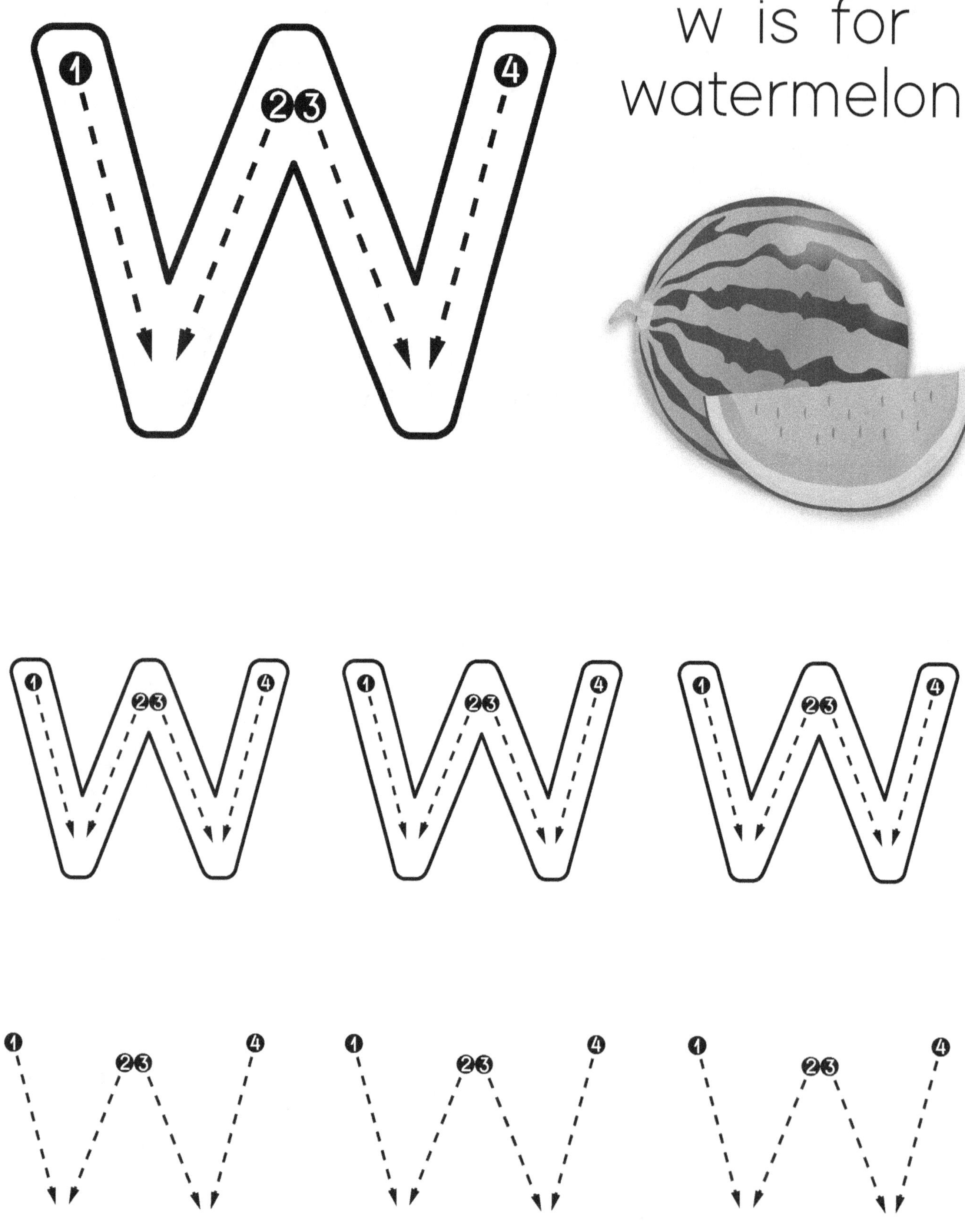

w is for
watermelon

Excellent!
Now
you can write

W

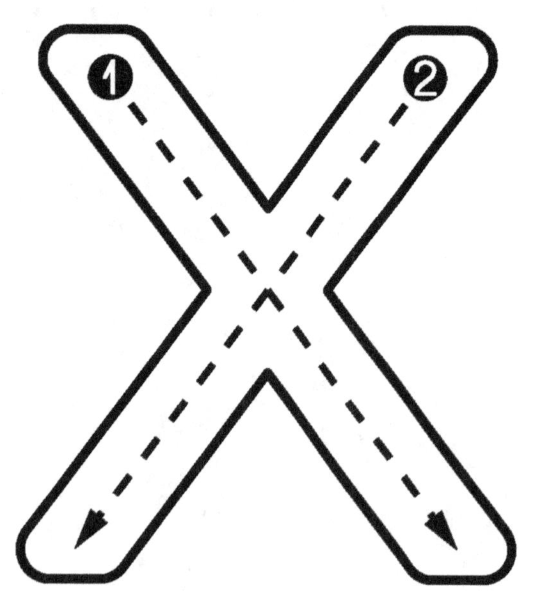

x is for
xylophone

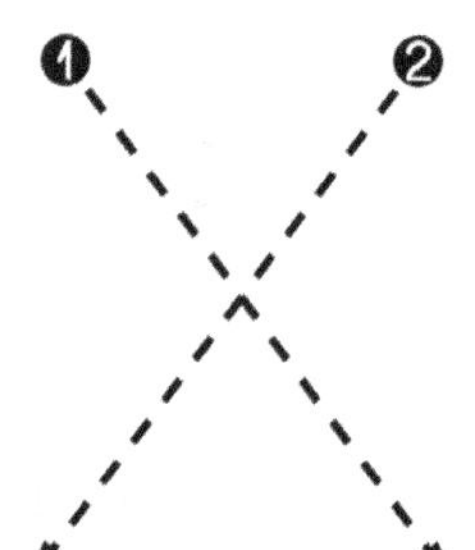 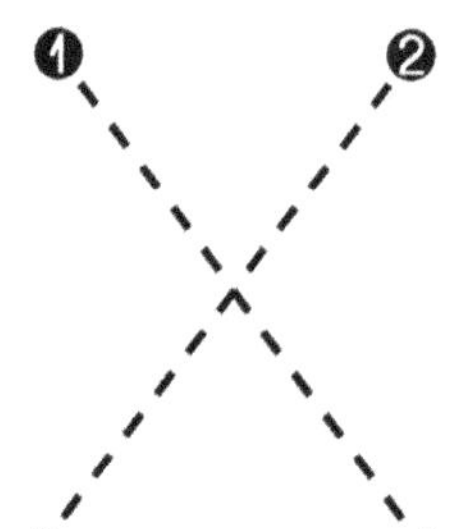 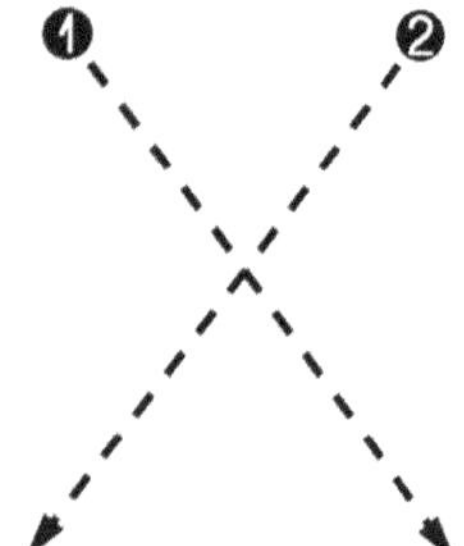

Excellent!
Now
you can write

X

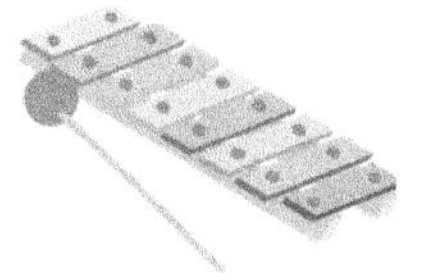

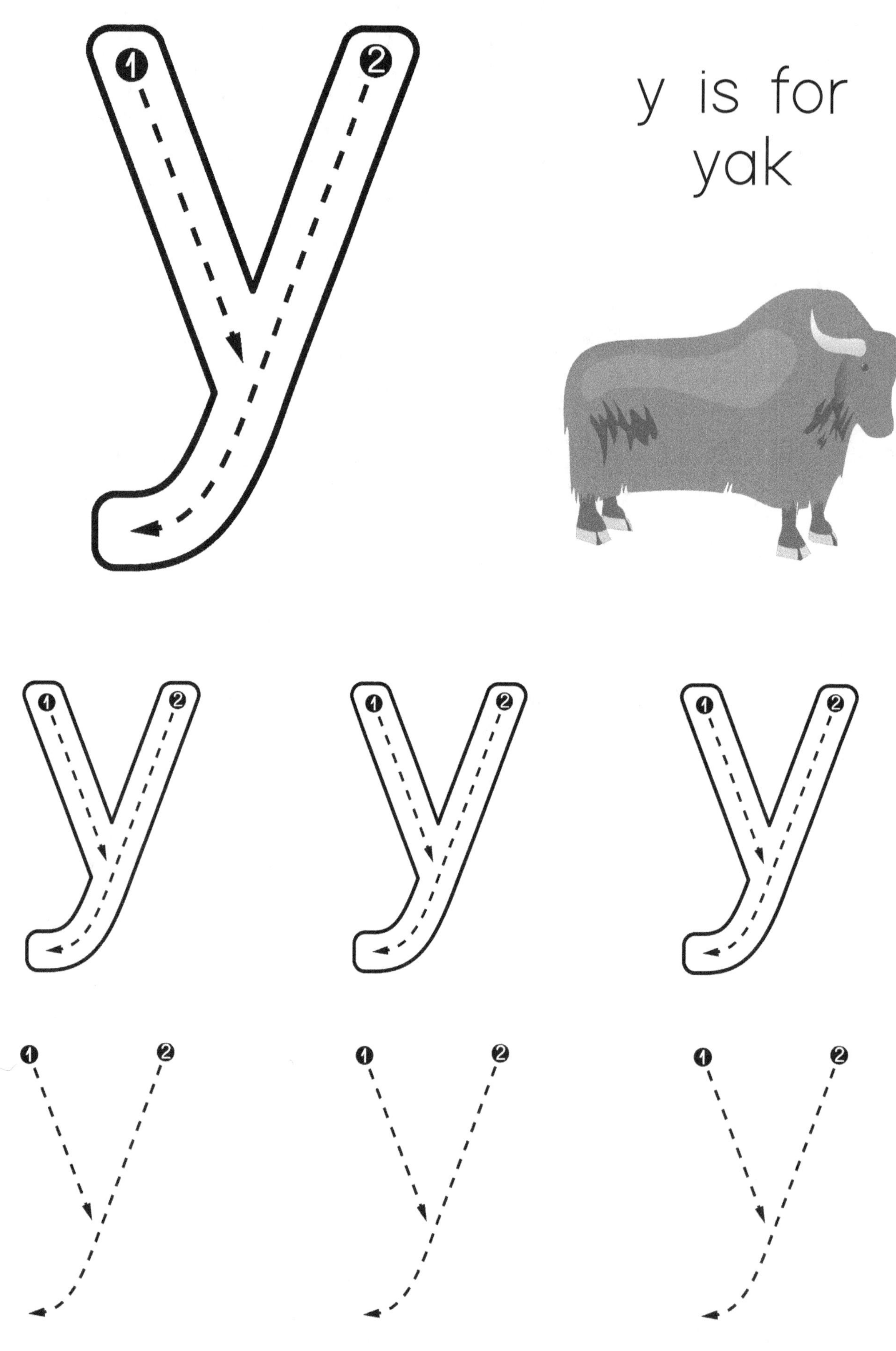

y is for
yak

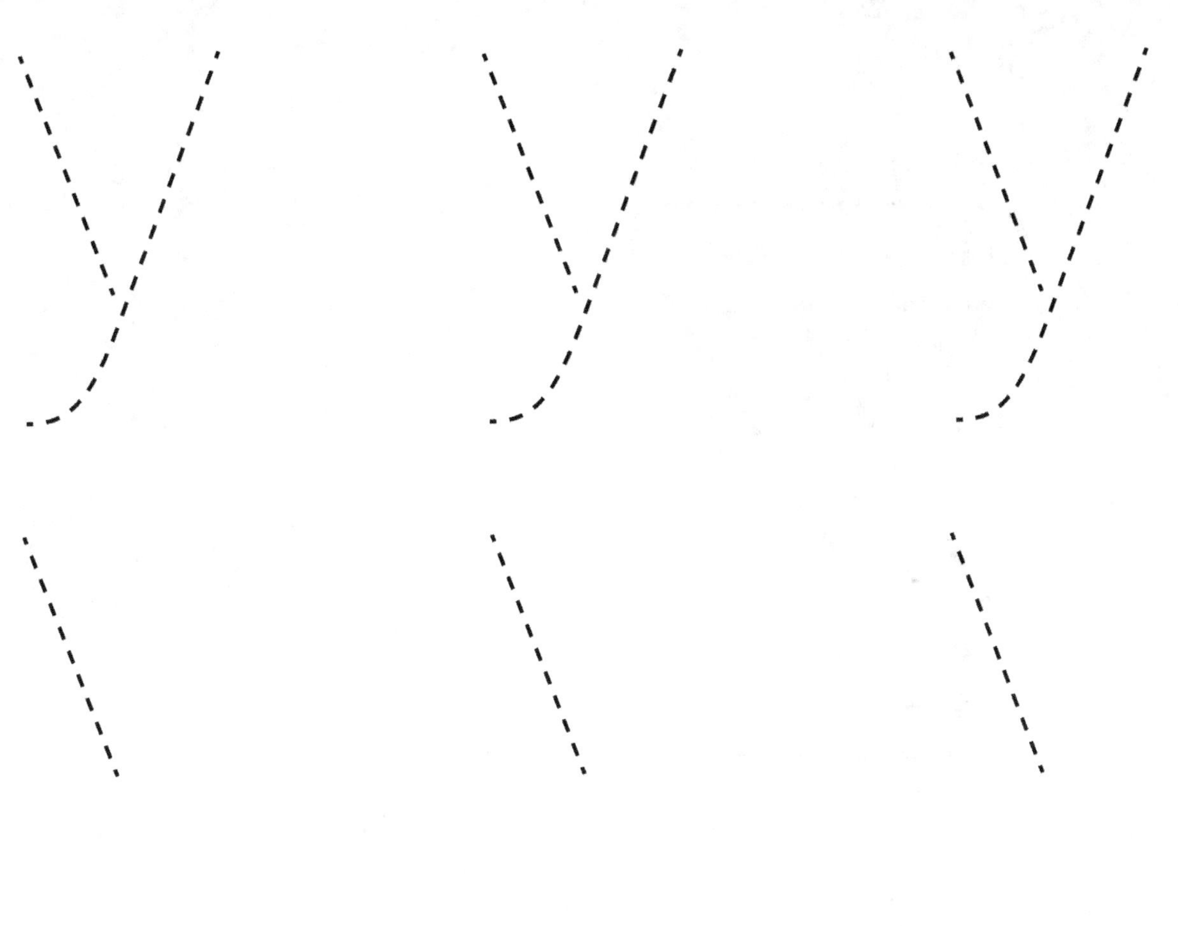

Excellent!
Now
you can write

y

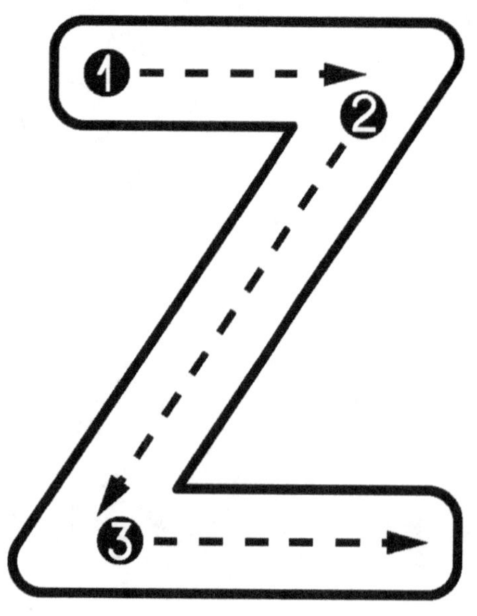

z is for
zebra

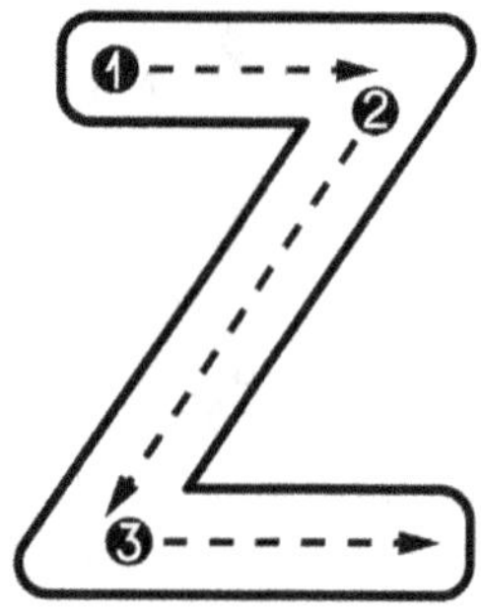

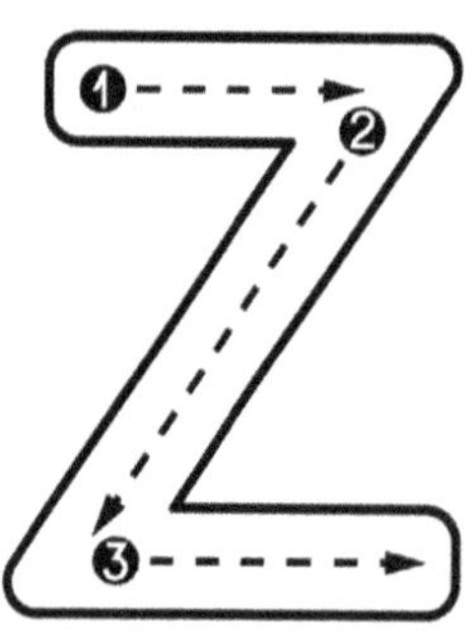

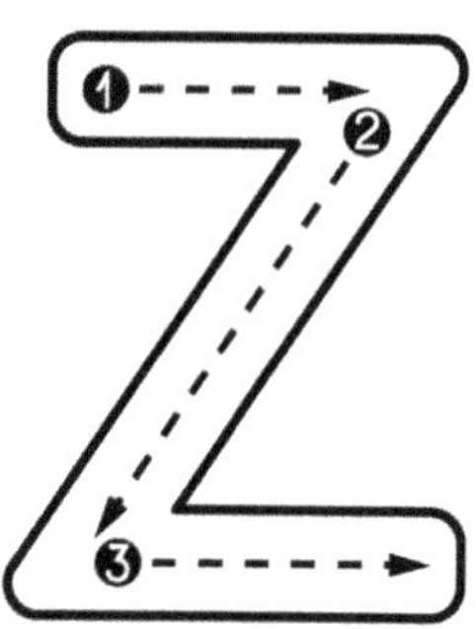

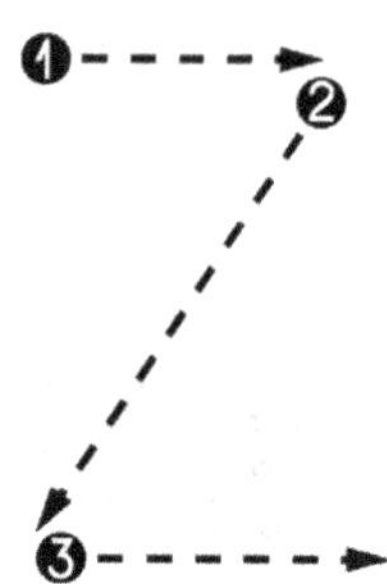

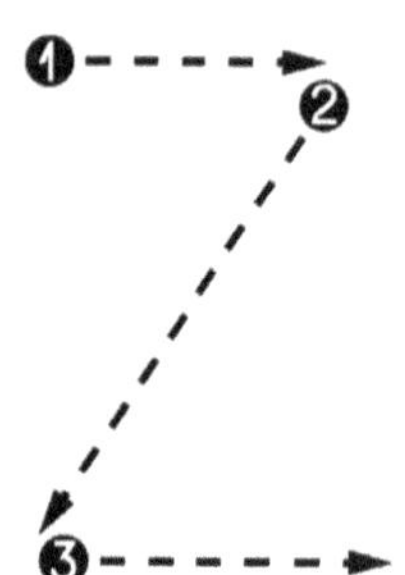

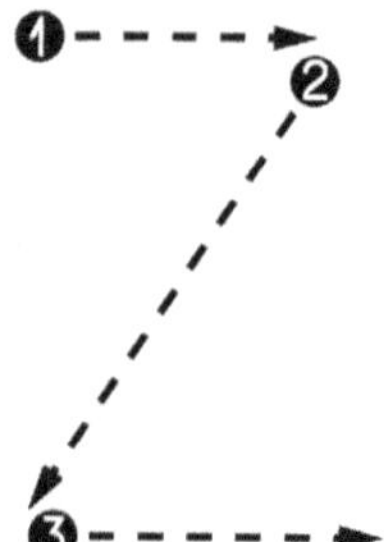

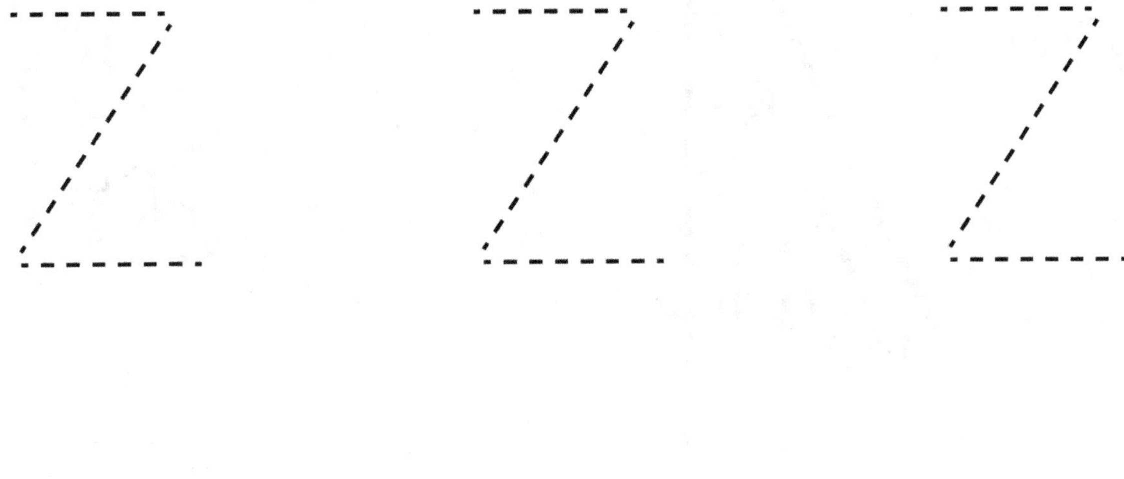

Excellent!
Now
you can write

z

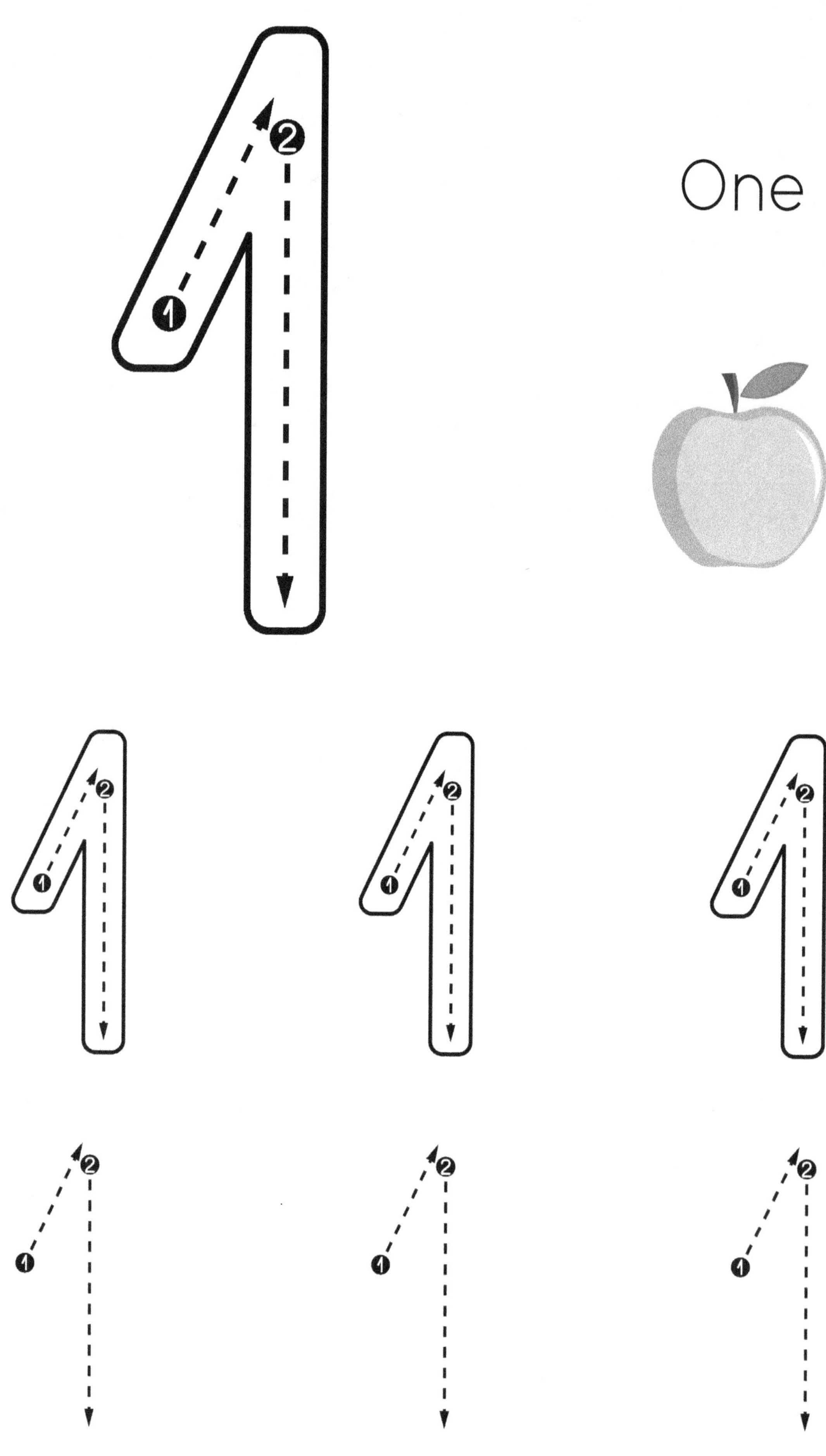
One

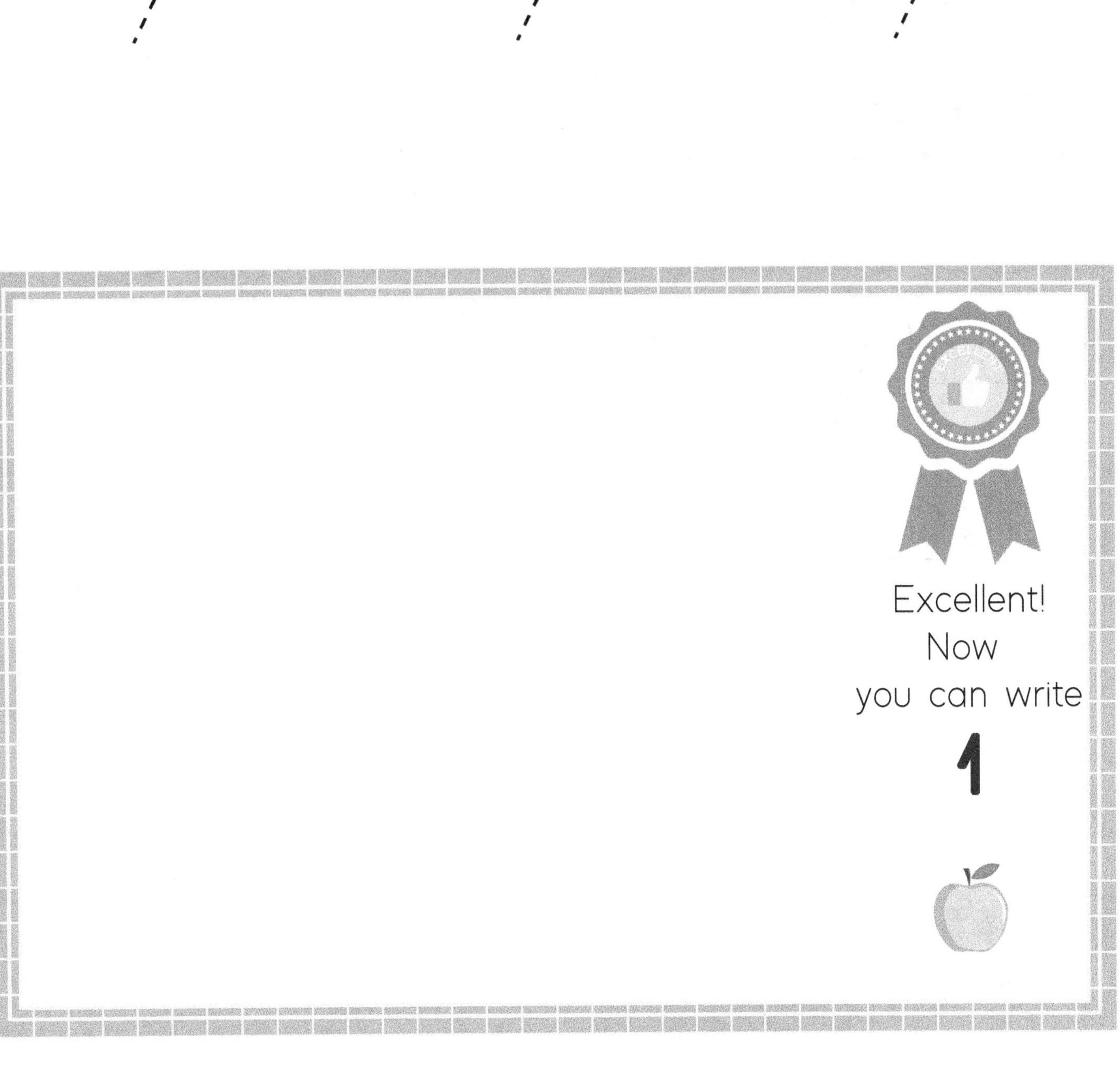
Excellent!
Now
you can write
1

Two

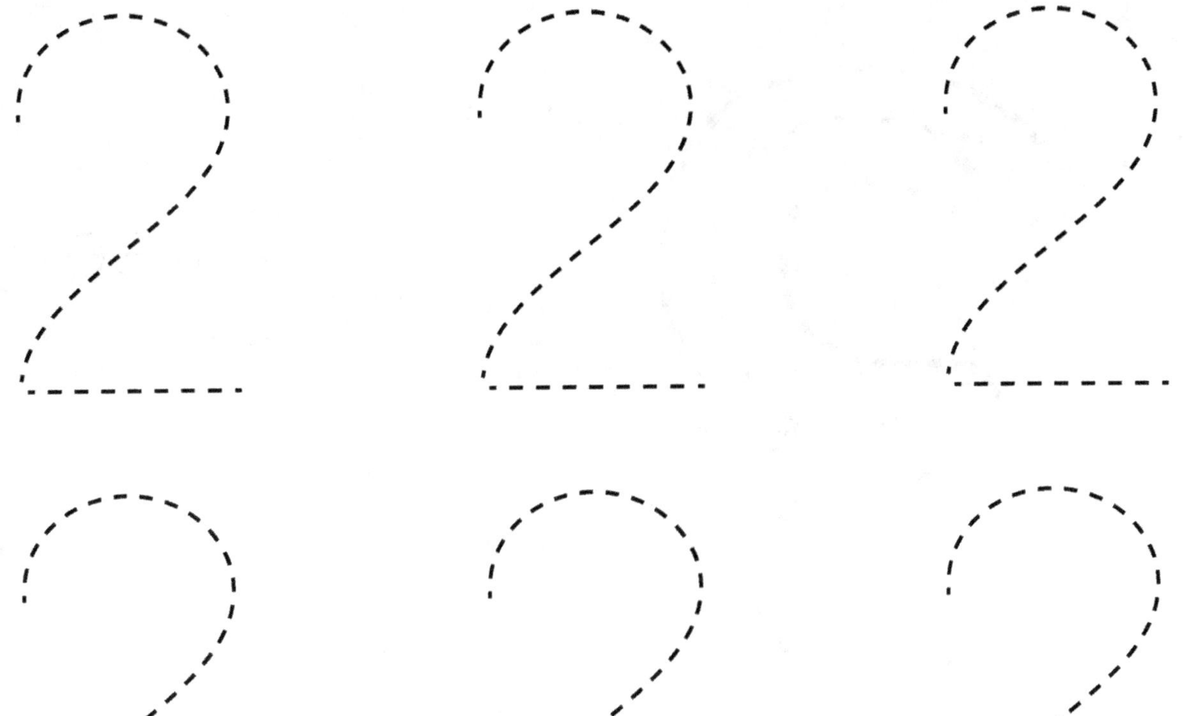

Excellent!
Now
you can write

2

Three

Excellent!
Now
you can write
3

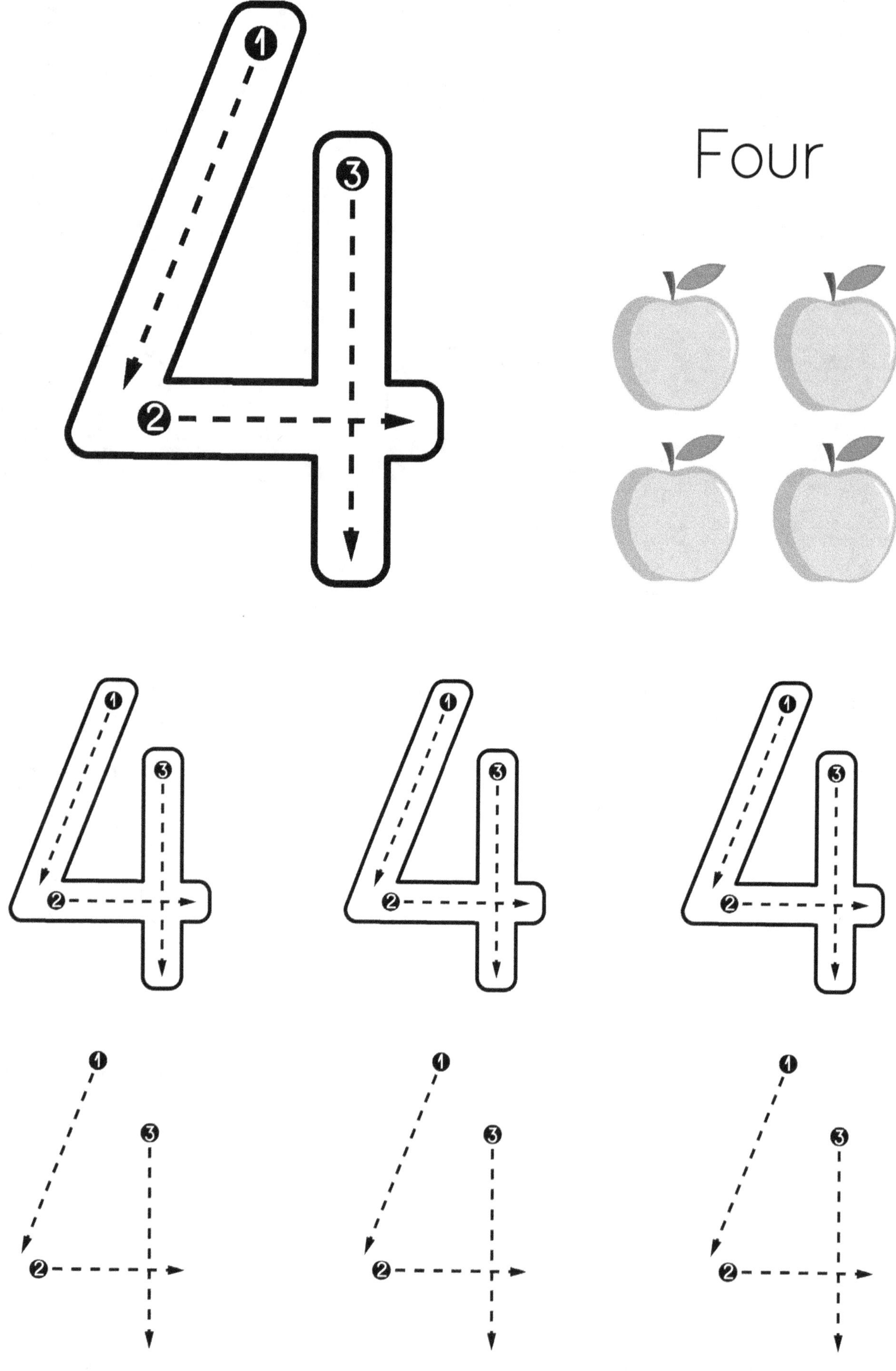

Four

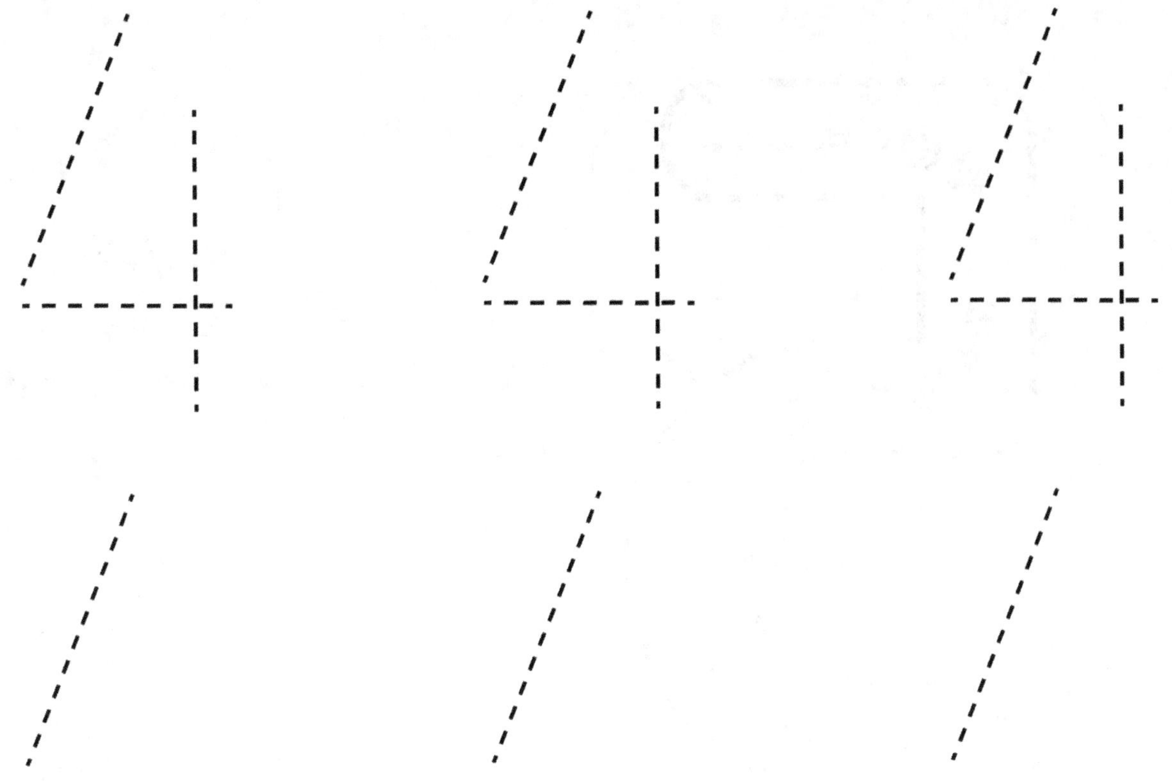

Excellent!
Now
you can write

4

Five

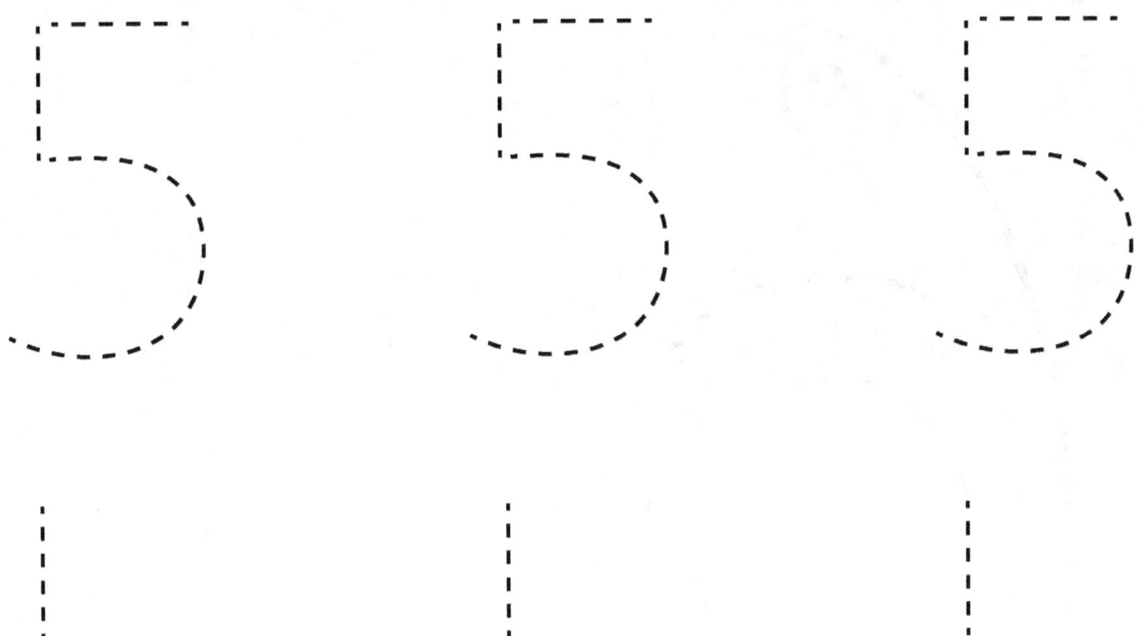

Excellent!
Now
you can write

5

Six

Excellent!
Now
you can write

6

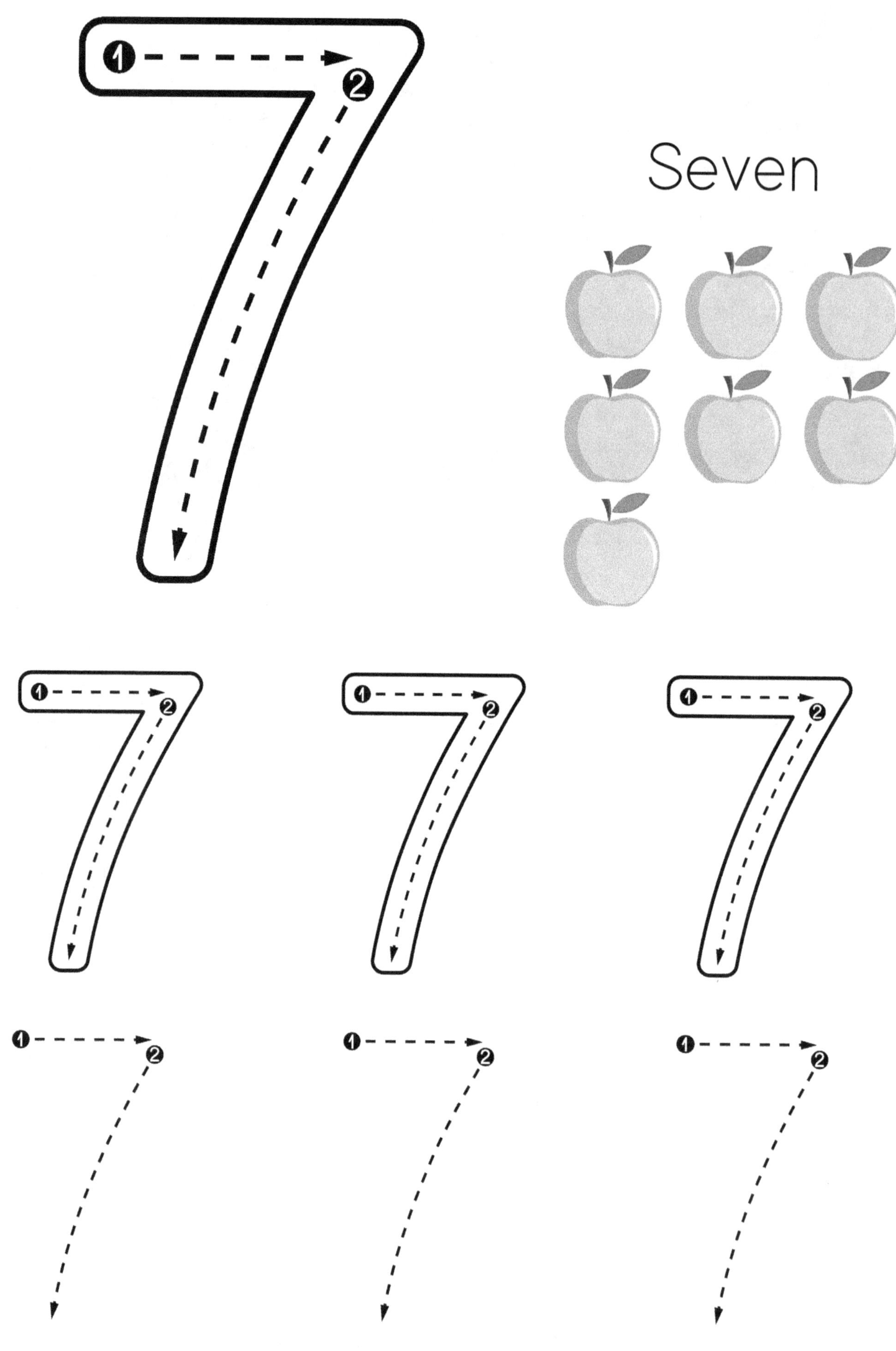

1
2
Seven

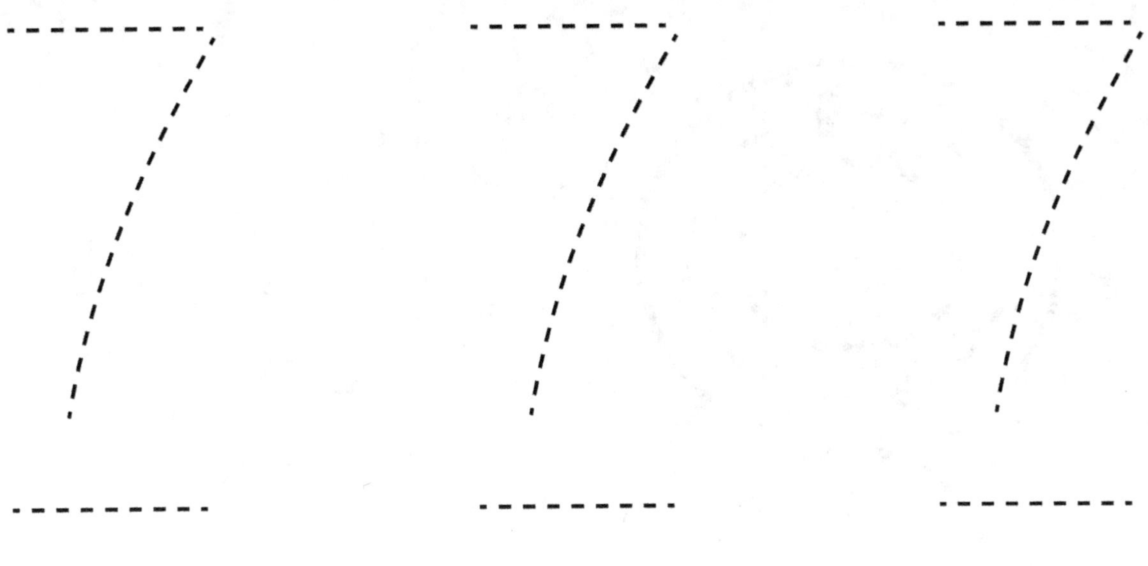

Excellent!
Now
you can write

7

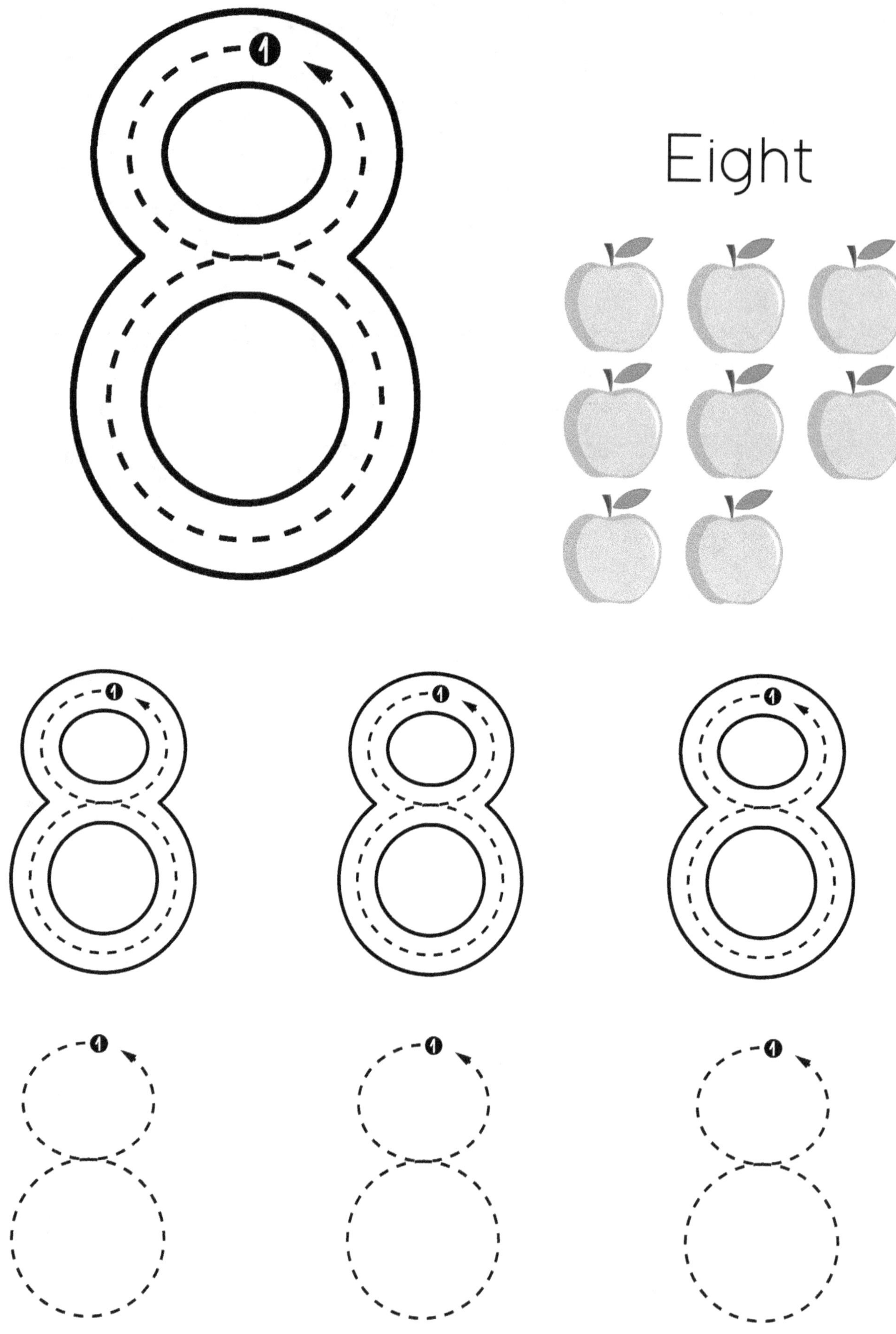

Eight

Excellent!
Now
you can write

8

Nine

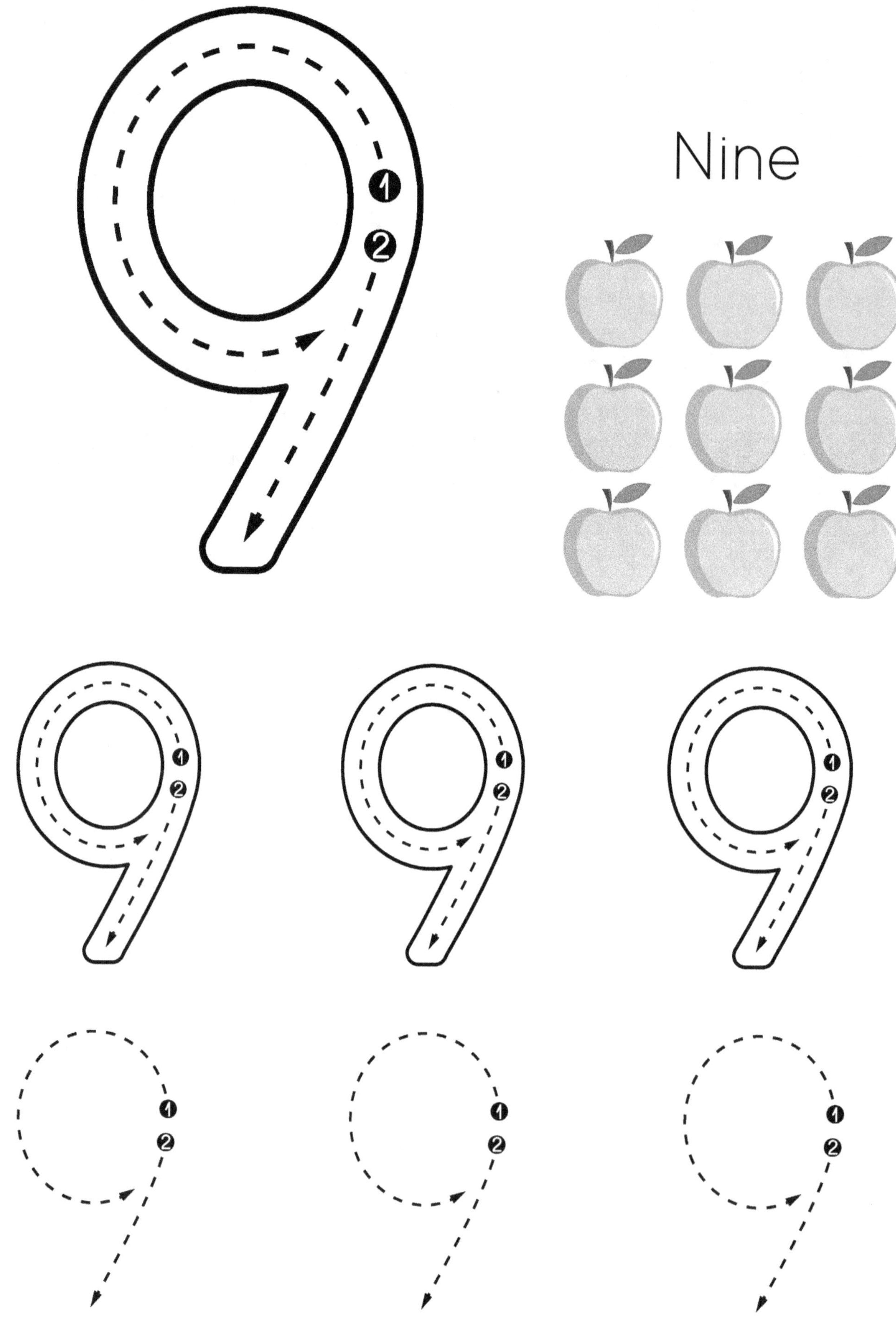

Excellent!
Now
you can write

9

Ten

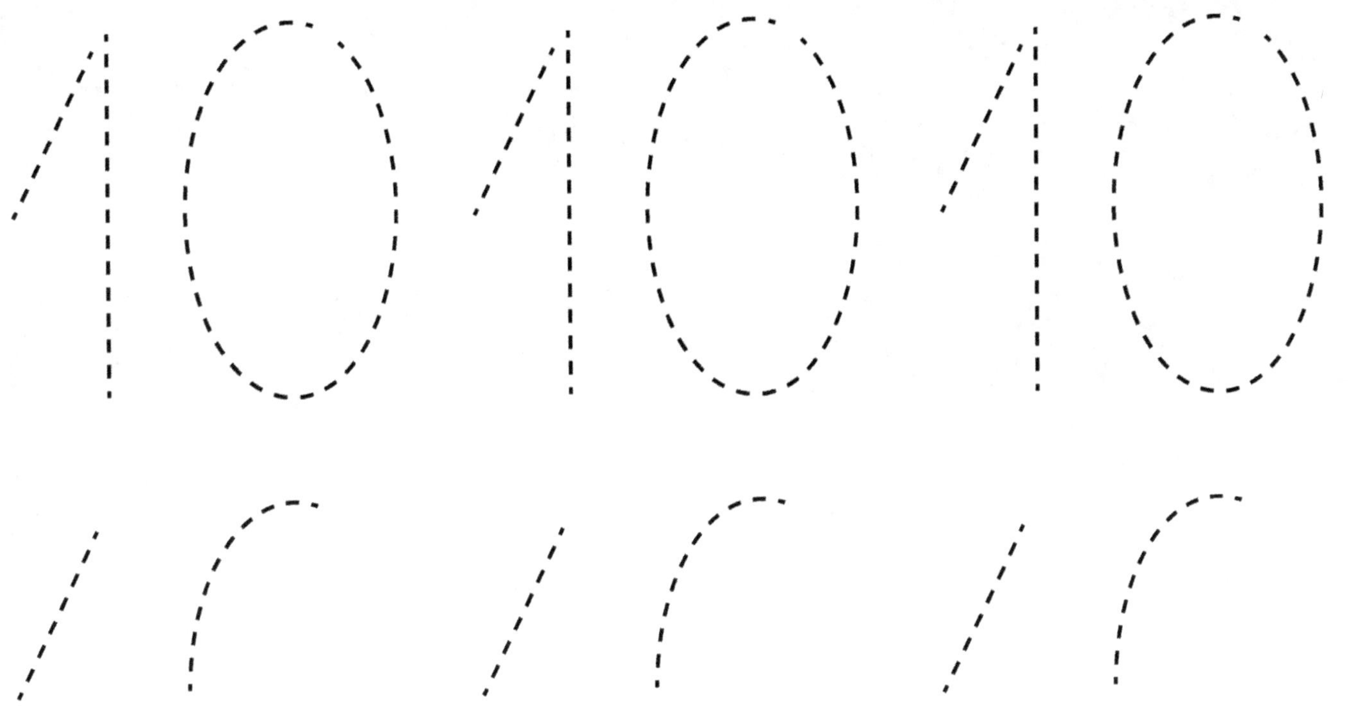

Excellent!
Now
you can write
10

How do you like our book?

We would really appreciate you leaving us a review.